Finite Mathematics
An Integrated Approach

Harold D. Shane
Baruch College
of CUNY

Charles E. Merrill Publishing Company
A Bell & Howell Company
Columbus, Ohio

MERRILL MATHEMATICS SERIES
Erwin Kleinfeld, Editor

Published by
Charles E. Merrill Publishing Company
A Bell & Howell Company
Columbus, Ohio 43216

ISBN: 0–675–08875–5

Library of Congress Catalog Card Number: 73-87111

2 3 4 5 6 7 8 9—80 79 78 77 76 75 74

PRINTED IN THE UNITED STATES OF AMERICA

To Erica and Lisa

Preface

This book is intended for a one-term, basic course. The prerequisite is a nodding acquaintance with high school algebra. Unlike most finite mathematics books, which treat the course as a collection of distinct noncalculus topics, this text attempts to integrate the major subjects in such a way as to achieve overall coherence. This integration is particularly noticeable in chapters 4 and 5, in which the matrix and vector algebra are defined to satisfy the needs in probability.

There are other features that are worth noting. The treatment of probability is done first without combinatorics (chapter 4) so that the concepts will not be obscured by the computational difficulties. We then return later to probabilistic problems at the end of chapter 7. Probability is discussed for a third time in chapter 6 as an application of the geometric progression.

The same kind of treatment is afforded to row operations on matrices which arise in the solution of systems of equations, again in the simplex procedure and for a third time in the evaluation of determinants. Similarly, linear programming appears twice, first in the context of lines in the plane, and again when the machinery of complete elimination has been developed. It is my hope that this helical approach, which brings us back repeatedly to the same concepts, will reinforce the student's understanding of the subject.

The motivation to write this text arose largely from the consistent dissatisfaction that most of our mathematics faculty showed with every text that was adopted for use in our basic course. I want to thank these people who are both my friends and my colleagues for their continual encouragement and advice. I am indebted to Selma Lippman, one of the few people able to read my handwriting, who spent many hours typing the manuscript. I also want to thank my wife Mita for solving all the exercises with me; she deserves the credit for the accuracy of the answers. Finally I would like to acknowledge the help, advice and encouragement given to me by Ian O'Leary, Pete Hutton, Maggie Shaffer and all the other fine people at Merrill who worked on this book.

Harold D. Shane

Contents

1

Sets

A set is a well-defined collection of objects, for example, the set of all presidents of the United States, past and present, or the set of whole numbers between 3 and 15, inclusive. Naturally, you would expect that most of the sets in a mathematics book would be sets of numbers. In this book, we shall talk a great deal about sets of numbers, but usually we shall talk about other types of sets. In particular, we shall speak of sets of points in the plane and sets of possible happenings when an experiment is performed. In any case, it is important to know how to describe a set so that it will be well defined. Perhaps the best way is to list the actual objects in the set. For example, consider the second set described above. For simplicity, let us call this set A. Now, $A = \{3, 4, 5, 6, 7, 8, 9, 10, 11, 12, 13, 14, 15\}$. Note that the curly brackets $\{\ \}$ are used to enclose all the numbers in A. By the way, the objects in the set are called *elements* or *points* of the set. We shall use the symbol $\in$ to indicate that an element belongs to a particular set. Thus $6 \in A$, and $17 \notin A$, where the / indicates "not," just as in the symbol $\neq$, for "not equal."

It is sometimes difficult to know whether a collection of objects is well defined and qualifies as a set. To begin with, one must have in mind some underlying universe from which points can be drawn. Thus, to say, "B is the set of all numbers larger than 2," presumes that one is thinking only of numbers which can be compared to 2 in terms of magnitude. In this case, the universe should consist of whole numbers or real numbers but not imaginary or complex numbers. Having decided upon a universe, one should now be able to take any element of that universe and decide whether or not the element is a member of the set. For example, let C be the set of all real numbers which are larger than $\sqrt{198}$. Now consider $7 + \sqrt{50}$: Is this number in C? To three decimal places, $\sqrt{198} = 14.071$ and $\sqrt{50} = 7.071$. So, with a limited table of square roots, one really does not know which is larger or if in fact they are equal. Of course, theoretically any two real numbers are comparable in terms of which is greater or smaller and so the set C is well defined. Nevertheless, there may be more to the innocuous sounding definition of a set than it would first appear. Actually, mathematicians, logicians and philosophers have devoted many years of effort to the problems involved in this fundamental notion. We shall not. In fact, we shall not discuss set theory at all. We shall, however, use the notation of sets extensively.

How else can we describe a set, other than by listing all its elements or writing a sentence about it? The most common method is to simply write the sentence in mathematical notation, such as $A = \{x : 3 \leq x \leq 15$, and x is an integer$\}$, read "A equals the set of numbers x such that x is greater than or equal to 3 and less than or equal to 15, and x is an integer." Of course, this is the same A described in the first paragraph. Note that ":" is read "such that." Having a set A and a known universe, one automatically can describe another set, namely, the set of all points not in A. We shall symbolize this by A' and call it "the complement of A". This is stated formally in the following definition.

Definition 1 $A' = \{x : x \notin A\} =$ Complement of A.

Suppose the universe is all the whole numbers between 1 and 100 inclusive. If $A = \{3, 4, 5, 6, 7, 8, 9, 10, 11, 12, 13, 14, 15\}$, then $A' = \{1, 2, 16, 17, 18, \ldots, 100\}$. The notation "..." means that the writer has left out something which he expects you to be able to fill in using common sense. Usually this is done to save writing; sometimes it is done because it cannot be avoided. For example, for $U = \{1, 2, 3, \ldots, 100\}$, the dots (or ellipsis) saves the writing of 96 numbers; for $V = \{2, 4, 6, \ldots\}$, V is the set of even positive integers and there is no largest element, so we cannot write them all out. A more complicated example would be $A_2 = \{1, 2\}$, $A_3 = \{1, 2, 3\}$, $A_4 = \{1, 2, 3, 4\}$, etc, or, in general, $A_n = \{1, 2, 3, \ldots, n\}$. We know that there is a largest element but we want to have a general expression which will be correct for all possible choices of n.

In the preceding example, it is clear that the elements of A_2 are included among the elements of A_4. When two sets are related in this way, the smaller is said to be a *subset* of the larger or we might say that the larger *contains* the smaller. This relationship is symbolized by $A_2 \subset A_4$. We shall state the following formal definition.

Definition 2 A is a subset of B if whenever $x \in A$, it is also true that $x \in B$.

Some immediate results of this definition are as follows:

1. For every set A, $A \subset A$.
2. If $A \subset B$ and $B \subset C$, then $A \subset C$.
3. If $A \subset B$ and $B \subset A$, then $A = B$.
4. If $A \subset B$, then $B' \subset A'$.

You should try to convince yourself that these are true although perhaps one comment is in order. We have written expressions of the form $A = B$ about two sets, but we have not said what is meant by equality of sets. Perhaps it would be simplest to say that Statement 3 is the definition of equality and let it go at that. But to be perfectly clear, let us say that two sets are equal if they contain exactly the same elements. Thus $\{1, 3, 2\} = \{2, 1, 3\} = \{3, 1, 2\}$, etc.

Suppose we have two sets, A and B. It is possible to form new sets by using the elements of both sets. For example, we define the intersection of A and B as the set of all points common to both sets. The symbol for this is $A \cap B$ and the formal definition follows.

Definition 3 $A \cap B = \{x : x \in A \text{ and also } x \in B\}$.

For example, if $A = \{1, 2, 3, \ldots, 10\}$, $B = \{6, 7, 8, \ldots, 15\}$ and $C = \{12, 13, 14, \ldots, 20\}$, then

$$A \cap B = \{6, 7, 8, 9, 10\}, \; B \cap C = \{12, 13, 14, 15\}$$

What about $A \cap C$? There are no points common to both A and C. We call such sets *disjoint*. Their intersection is *empty*, that is, it is a set with no elements in it. Such a set is called the *empty set* or *null set* and is identified by the symbol $\varnothing$. Null sets usually arise in one of two ways. The first we have already seen, that is, as the intersection of disjoint sets. The second occurs when a set is described by a mathematical condition which can never be true. For example, $\{x : x^2 = -1$ and x is a real number$\}$ $= \varnothing$. Certain properties of intersections now follow immediately from Definitions 2 and 3:

5. $A \cap B = B \cap A$.
6. $A \cap A = A$.
7. $A \cap U = A$, where U is the universe.
8. $A \cap \varnothing = \varnothing$.
9. $A \cap B \subset A$, and $A \cap B \subset B$.
10. $\varnothing \subset A$, for every set A.

Let us consider another operation involving two sets. The union of A and B is the set of all points in one or the other of the sets or in both. This operation is symbolized $A \cup B$. Thus we have the following Definition.

Definition 4 $A \cup B = \{x : x \in A$ or $x \in B$ or both$\}$.

So, using A, B and C as in the last example, $A \cup B = \{1, 2, 3, \ldots, 15\}$ and $B \cup C = \{6, 7, 8, \ldots, 20\}$. Notice that points which occur in both sets are not written twice in the union. This is so because a point is either in a set or not in a set and indicating it more than once would serve no useful purpose.

Before we list some further properties of complementation, union and intersection, a few remarks about the order of operations should be made. First, the complementation symbol applies only to the set to which it is attached. That is, $A \cap B'$ is the intersection of A with the complement of B. To indicate the complement of the intersection of A and B, one would write $(A \cap B)'$. Here, just as in algebra, the operation within the parentheses is to be performed first. The symbol $A \cap B \cup C$ is unclear, since no agreement has been reached as to which takes precedence, union or intersection. In fact, we shall not reach an agreement, but we will use parentheses in either case. That is, $A \cap (B \cup C) \neq (A \cap B) \cup C$, but at least both sides have a unique interpretation. In order to deduce some of the properties of sets, it is convenient to use a device known as the Venn diagram. In a Venn diagram, the interior of a square is taken to be the universe. Regions bounded by given closed curves are the sets in this universe and intersections and unions have obvious geometric representations. Figure 1 shows Venn diagrams for a number of typical sets which may be encountered. In particular, 1c shows that $A \cup B$ is the union of three disjoint sets, $A \cap B'$, $A \cap B$ and $A' \cap B$. The idea that a set can be partitioned in this way, into a so-called disjoint union of subsets, is quite useful in probability problems.

Let us now list some additional simple properties of sets.

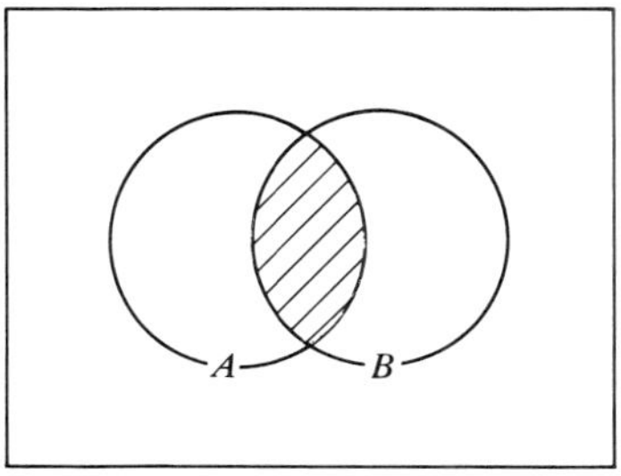

(a)
$A \cap B$ is shaded.

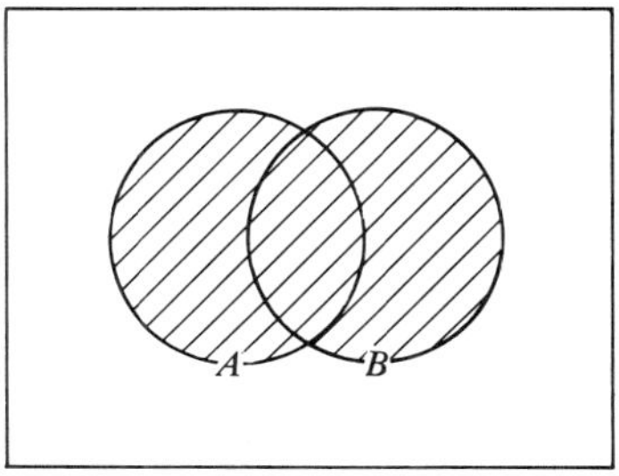

(b)
$A \cup B$ is shaded.

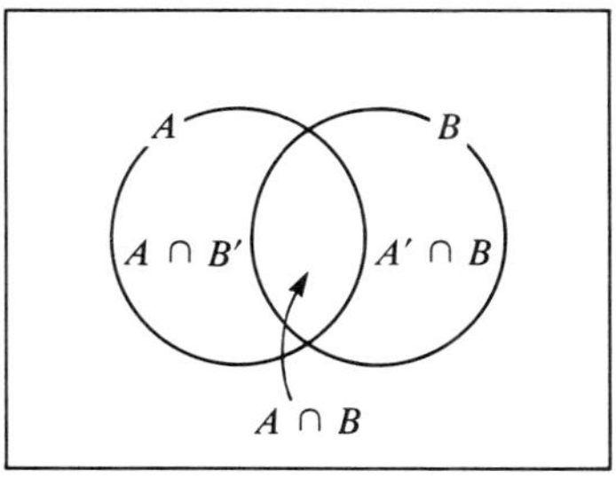

(c)
$A \cup B = (A \cap B') \cup (A \cap B) \cup (A' \cap B)$

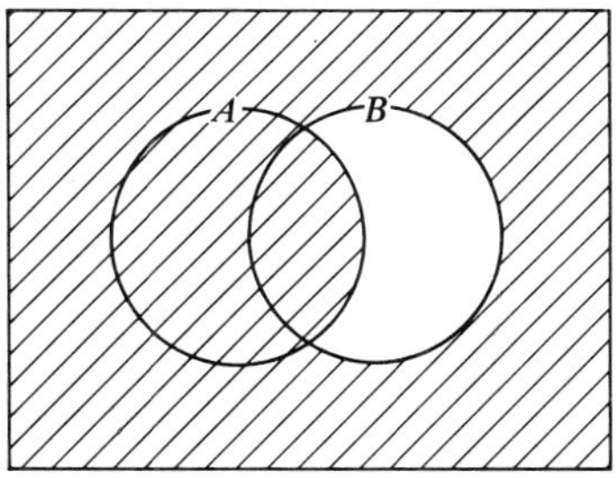

(d)
$A \cup B'$ is shaded.

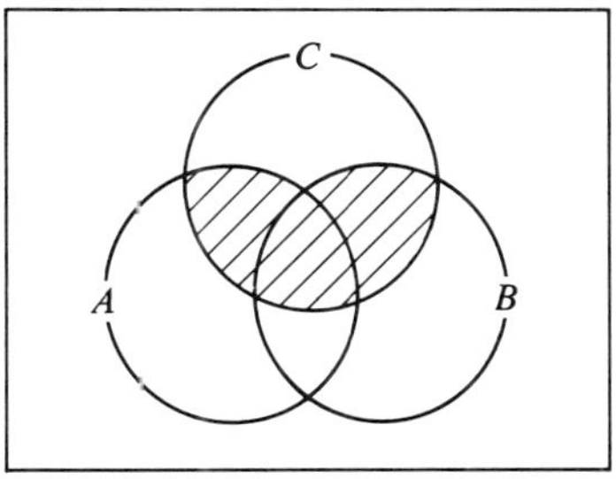

(e)
$(A \cup B) \cap C$ is shaded.

Figure 1

For unions:

11. $A \cup B = B \cup A$.
12. $A \cup A = A$.
13. $A \cup U = U$, where U is the universe.
14. $A \cup \varnothing = A$.
15. $A \subset A \cup B$, and $B \subset A \cup B$.

For complementation:

16. $(A')' = A$.
17. $U' = \varnothing$, and $\varnothing' = U$.

Finally, some less than obvious results:

18. $A \cap (B \cap C) = (A \cap B) \cap C$, and $A \cup (B \cup C) = (A \cup B) \cup C$, the associative law (which allows us to drop parentheses in cases of multiple union or intersection).
19. $A \cap (B \cup C) = (A \cap B) \cup (A \cap C)$, and $A \cup (B \cap C) = (A \cup B) \cap (A \cup C)$, the distributive laws.
20. $(A \cup B)' = A' \cap B'$, and $(A \cap B)' = A' \cup B'$, DeMorgan's laws.

The last two are not easy to see and you should draw Venn diagrams to convince yourself that they are at least plausible.

EXERCISES

1. Given the universe $U = \{-1, 0, 1, 2, 3, 4, 5, 6, 7\}$, $R = \{1, 2, 3, 4\}$, $S = \{-1, 0, 1, 2\}$, $T = \{2, 4, 6\}$, and $X = \{1, 3, 5, 7\}$. Which of the following are true?
   (a) $X \cap T = \varnothing$
   (b) $X \cup T = U$
   (c) $T \cap S \subset R$
   (d) $T \cap S' \subset X$
   (e) $X \cap T' = X$
   (f) $(X \cap T)' = U$
   (g) $R \subset S' \cup T'$
   (h) $S' \subset X \cup T$

In Exercises 2–7, the universe $U = \{1, 2, 3, \ldots, 50\}$, $A = \{1, 2, 3, \ldots, 25\}$, $B = \{10, 11, 12, \ldots, 35\}$ and $C = \{2, 4, 6, \ldots, 50\}$.

2. Find A', B' and C'.
3. Find $A \cap B$, $A \cap C$, $A' \cap C$ and $A' \cap B' \cap C'$.
4. Find $A \cup B$, $A \cup C$, $A' \cup C$, $C' \cup B'$ and $B \cup (A \cap C)$.
5. Find $A \cap (B \cup C)$, $(A \cap B) \cup (A \cap C)$, and verify that the distributive law holds.
6. Repeat Exercise 5 for $A \cup (B \cap C)$ and $(A \cup B) \cap (A \cup C)$.
7. Verify DeMorgan's laws: $(A \cup B)' = A' \cap B'$, and $(A \cap B)' = A' \cup B'$.
8. Construct Venn diagrams to verify the distributive laws and DeMorgan's laws.
9. Let $D = \{-2, -1, 0, 1, 2, 3, 4\}$; find $R = \{y : y = x^2, \text{ for } x \in D\}$.
10. Let $R = \{1, 2, 3, 4, 5, 6\}$; find $D = \{x : x^2 = y, \text{ for } y \in R\}$.
11. Let $D = \{-1, 0, 1, 2, 3, 4, 5\}$; find $E = \{y : x + (1/2)y = 3, \text{ for } x \in D\}$.
12. Find $F = \{(x, y) : x + 2y = 15 \text{ and } x, y \text{ are positive integers}\}$.
13. List all the possible subsets of $H = \{\#\}$. Do the same for $S = \{@, \$\}$ and for $T = \{\%, ¢, \&\}$. (Do not forget the empty set and the set itself!) Can you guess a rule that tells you the number of subsets of a set with n elements?
14. Let A be the set of students who passed the first examination and B be the set of students who passed the second examination. A consists of 25 students, B consists of 18 students and $A \cap B$ contains 13 students. How many students are there in $A \cup B$? How would you describe the sets $A \cap B$ and $A \cup B$ in words?
15. In Exercise 14, suppose that there are 25 students in A, 23 students in B and 29 students in $A \cup B$. How many students passed both exams?

2

Subsets of the Plane

2.1 Linear Equations

In this chapter, we shall deal with a universe which consists of all ordered pairs of real numbers. That is, $U = \{(x, y) : x \text{ and } y \text{ are any real numbers}\}$. So, for example, $(1, 4)$, $(-1/2, 5)$, $(\sqrt{2}, -\pi)$ are all examples of points in U. Since order counts, the points $(1, 2)$ and $(2, 1)$ are two distinct elements of U. As you know, when coordinate axes are introduced in the plane, every element of U is associated with a point in the plane and, conversely, every point in the plane has attached to it a unique ordered pair of coordinates. In case you have forgotten, let us briefly review how this is done. First, as in figure 1, a pair of number lines are drawn at right angles to one another, intersecting at the point zero on each line.

The horizontal line is called the x-axis and the vertical line is called the y-axis. Now through any point in the plane, construct a vertical line. At the point where this line crosses the x-axis is a number called the x-coordinate (or abscissa) of the point. Now construct a horizontal line through the point. At the point where this line crosses the y-axis is a number called the y-coordinate (or ordinate) of the point. If the point is called P, and its "coordinates" are (x, y), then we shall sometimes refer to it as $P(x, y)$. In figure 1, the points $P(1, 2)$, $Q(-1, 1/2)$ and $R(-2, -2\frac{1}{4})$ are shown.

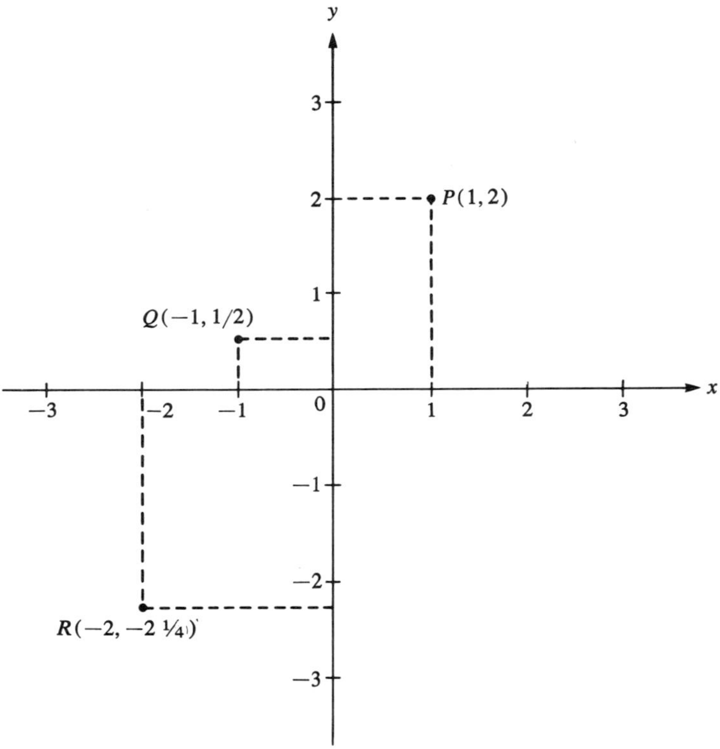

Figure 1

A set in this universe is some collection of points in the plane, which may be described as $\{(x, y) : \text{some relationship between } x \text{ and } y \text{ is true}\}$. For example, $L = \{(x, y) : x = 3\}$ is the set of all points having x-coordinate 3. Geometrically speaking, this would be a vertical line crossing the x-axis at 3. Actually, we should not talk about the point 3. Every point must have two coordinates in order to identify it, so the point at x-value 3 on the x-axis has coordinates $(3, 0)$. In fact, every point on the x-axis has ordinate 0 and every point on the y-axis has abscissa 0. The point where the axes intersect, called the *origin*, has coordinates (0, 0).

Returning to our example, if, in general, $L = \{(x, y) : x = a\}$, then geometrically L is a vertical line passing through $(a, 0)$, assuming, of course, that a is a constant. Usually we will bypass the set notation and just refer to "the line $x = a$." Similarly, "the horizontal line $y = b$" is short for $K = \{(x, y) : y = b\}$. Incidentally, the set $K \cap L = \{(a, b)\}$, that is, the geometric intersection of the two lines and the set intersection are one and the same!

Let us now consider some lines not parallel to either axis. To begin, we take a line L_1 passing through the origin and making an angle α with the positive x-axis (see figure 2). If $P(x, y)$ is any point on the line, other than the origin, then $\tan \alpha = y/x$. Using the symbol $m = \tan \alpha$, we obtain $m = y/x$ or $y = mx$ as the equation of the line. If we consider another line L_2, parallel to L_1 and cutting the y-axis at $(0, b)$, then for each x-value the y-value on L_2 will be simply b more than the value on L_1 (see figure 3). That is, the equation of this line will be $y = mx + b$. In this equation, m is called the slope of the line and b is called the y-intercept. It should be obvious that if two nonvertical lines are parallel, then they have the same slope. Figure 4 shows some typical graphs of lines of the form $y = mx + b$. How does one draw the graph of such a line?

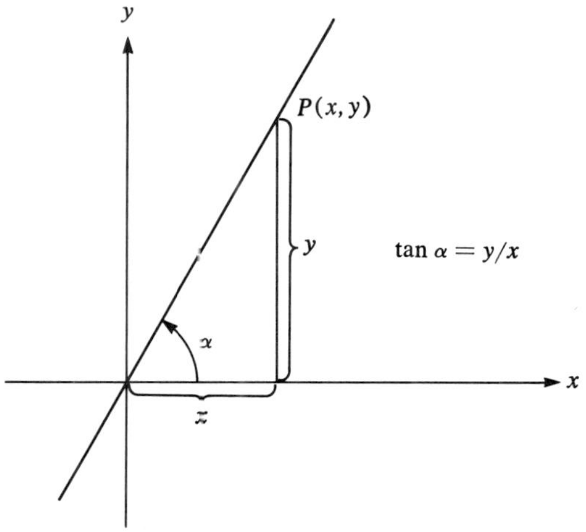

Figure 2

Example 1 Graph the straight line whose equation is $y = -2x + 4$.

According to Euclid, two points determine a straight line. Hence, we need only substitute two values of x, find the corresponding values of y, plot

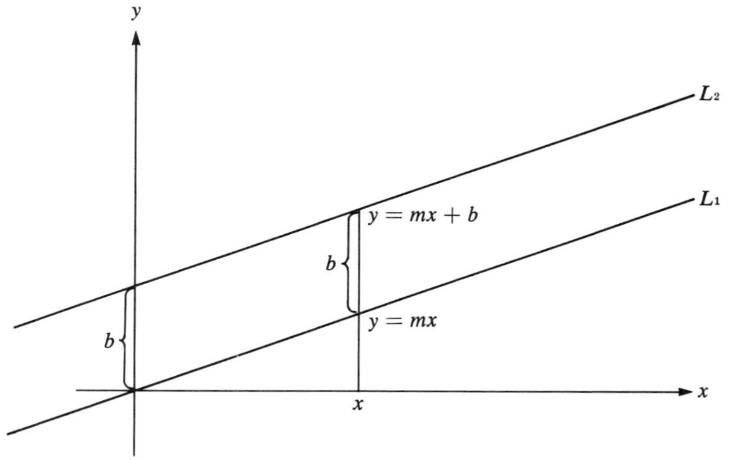

Figure 3

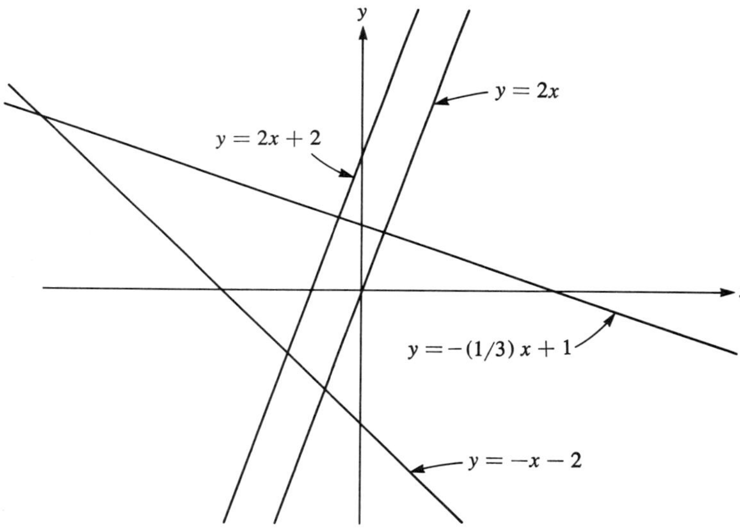

Figure 4

these two points and draw the line. So when $x = 0$, $y = 4$; and when $x = 1$, $y = 2$. Thus, (0, 4) and (1, 2) lie on the line and the graph is shown in figure 5.

Example 2 Find the equation of a line of slope 3, which passes through the point $(-1, 4)$.

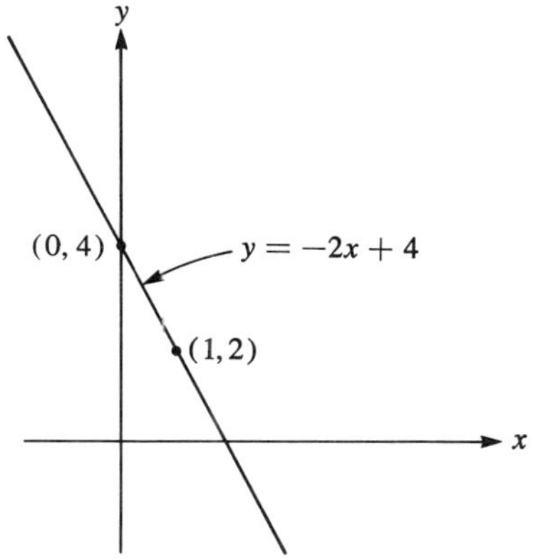

Figure 5

The equation of the line must be $y = 3x + b$, since it has slope 3. In addition, the equation must be true for $x = -1$ and $y = 4$. That is,

$$4 = 3(-1) + b$$
$$4 = -3 + b \qquad b = 7$$

Hence, the required equation is $y = 3x + 7$.

Example 3 Find the equation of the line which passes through (2, 9) and (−5, −6).

The equation is $y = mx + b$, which must be true at both points. Therefore,

$$9 = 2m + b$$
$$-6 = -5m + b$$

Here we have two equations in two unknowns, m and b. To solve, we subtract the lower equation from the upper, which yields

$$15 = 2m - (-5m)$$
$$15 = 7m$$
$$m = 15/7$$

Substituting into the first equation, we get

$$9 = 2(15/7) + b$$
$$9 = 30/7 + b$$
$$b = 9 - 30/7 = 63/7 - 30/7 = 33/7$$

Hence, the required equation is

$$y = (15/7)x + 33/7$$

which looks nicer if both sides are multiplied by 7 to yield

$$7y = 15x + 33$$

In Example 3, we see that given two points on a line, we are able to determine the slope of the line. Now, let us try to do this in general. Suppose we know that a line passes through two points $P(x_1, y_1)$ and $Q(x_2, y_2)$. Following the pattern in the example,

$$\begin{aligned} y_2 &= mx_2 + b \\ y_1 &= mx_1 + b \\ \hline y_2 - y_1 &= mx_2 - mx_1 \\ y_2 - y_1 &= m(x_2 - x_1) \\ m &= \frac{y_2 - y_1}{x_2 - x_1} \quad \text{if } x_2 \neq x_1 \end{aligned}$$

This is called the *slope formula*. Of course, if P and Q are different points and $x_2 = x_1$, the line is vertical and the slope is undefined.

Example 4 Use the slope formula to solve the problem in Example 3.

In this case $x_1 = 2$, $y_1 = 9$, $x_2 = -5$ and $y_2 = -6$, so

$$m = \frac{-6 - 9}{-5 - 2} = \frac{15}{7}$$

Therefore,

$$y = (15/7)x + b$$

and using the point (2, 9),

$$\begin{aligned} 9 &= (15/7)2 + b \\ b &= 33/7 \end{aligned}$$

and $y = (15/7)x + 33/7$, as before.

In Examples 2 and 4, we were faced with the problem of knowing the slope of a line and one point on the line and needing to know its equation. Using the slope formula, you will find that it is easy to find a general solution to this problem. Let L be a line with slope m, passing through the point $P(x_1, y_1)$. Let (x, y) be any other point on the line. Now, using the slope formula with (x, y) playing the role of Q, we have

$$m = \frac{y - y_1}{x - x_1}$$

that is,

$$y - y_1 = m(x - x_1)$$

This is called the point-slope formula for a straight line.

Example 5 Using the point-slope formula, rework Example 4.

We have $m = 15/7$ and $(x_1, y_1) = (2, 9)$. Hence the equation is

$$y - 9 = (15/7)(x - 2)$$

Multiplying both sides by 7, we have

$$\begin{aligned} 7y - 63 &= 15(x - 2) \\ 7y - 63 &= 15x - 30 \\ 7y &= 15x + 33 \end{aligned}$$

which is the same equation obtained previously.

The slope formula also gives us a good indication of what the slope really represents. If we always think of Q as lying to the right of P, then the denominator, $x_2 - x_1$, is always positive. If Q is higher than P, the numerator is also positive; hence, m is positive. So if the line is rising as you go from left to right, the slope is positive; otherwise it is negative. A zero slope indicates that the numerator is zero, which means that $y_2 = y_1$ and the line is horizontal. A line which makes a 45° angle with the x-axis has slope $+1$ (-1 means 45° below the positive x-axis). Large slopes indicate steepness and small slopes indicate a gentle rise or fall.

We have seen that the equation of a line may take one of three forms:

1. $x = a$, a vertical straight line,
2. $y = b$, a horizontal straight line,
3. $y = mx + b$, any other line.

Of course, Form 2 is only a special case of Form 3 with $m = 0$. All this can be summarized in the following.

THEOREM 1 Every equation of the form $Ax + By = C$, A and B not both zero, is the equation of a straight line. A, B and C are constants. (Accordingly, every such equation is called a *linear equation.*)

Proof If $A = 0$, $B \neq 0$, $By = C$, $y = C/B$, which is a horizontal line. If $B = 0$, $A \neq 0$, $Ax = C$, $x = C/A$, which is a vertical line. If neither A nor B is zero,

$$\begin{aligned} Ax + By &= C \\ By &= -Ax + C \\ y &= (-A/B)x + C/B \end{aligned}$$

which is of the form $y = mx + b$.

Example 6 Find the slope and y-intercept of the line, with the equation $3x + 4y = 8$. Find the x-intercept, too. (The x-intercept is the point where the line crosses the x-axis.)

The equation is $3x + 4y = 8$, which we solve for y:

$$\begin{aligned} 4y &= -3x + 8 \\ y &= (-3/4)x + 2 \end{aligned}$$

Therefore, the slope is $-3/4$ and the y-intercept is $(0, 2)$. To find the x-intercept, we must find the value of x for which $y = 0$. In other words, substitute $y = 0$ in the equation and solve for x.

$$\begin{aligned} 3x + 4(0) &= 8 \\ 3x &= 8 \qquad x = 8/3 \end{aligned}$$

So the x-intercept is $(8/3, 0)$.

In general, unless a line passes through the origin, the easiest way to draw the graph of a line is to plot its x- and y-intercepts.

Example 7 Plot the line $3x + 5y = 10$.

Let $x = 0$. We have $5y = 10$, $y = 2$, that is, the y-intercept is 2. Let $y = 0$. We have $3x = 10$, $x = 10/3$; the x-intercept is $10/3$. The graph is shown in figure 6.

EXERCISES

1. Find the equations of the lines whose slopes and y-intercepts are
   (a) $2, 4$
   (b) $-3, 0$
   (c) $0, -2$
   (d) $1/2, -1/2$
   (e) $-1/4, 3$

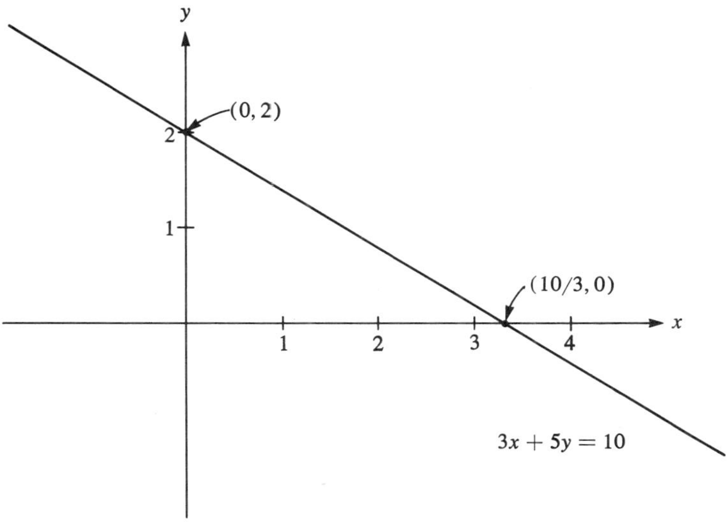

Figure 6

2. Find the equations of the lines with the given slopes and passing through the given points:
   (a) -3, (2, 5)
   (b) 1/2, $(-2, 0)$
   (c) 1/6, (0, 0)
   (d) $-1/4$, $(2, -2)$
3. Find the slope, x- and y-intercepts of $2x - 5y = 6$. Plot the line.
4. Find the equation of the line parallel to the line in Exercise 3 passing through $(-1, 7)$. Plot the line.
5. Plot the line $2x + 6y + 11 = 0$. Find the area of the triangle formed by the line, the x-axis and the y-axis.
6. Find the equation of the vertical line passing through the x-intercept of the line in Exercise 5.
7. Find the equations of the two lines parallel to $y = -3$, and 4 units away from it.
8. Find the equations of the line passing through the given pair of points:
   (a) $(1/2, -2)$ and $(1/4, -1/4)$
   (b) $(0, -9)$ and $(-1/2, 3)$
   (c) $(-1/3, 2/3)$ and $(0, 0)$
   (d) $(1, -5)$ and $(2, -5)$
   (e) $(1/2, 16)$ and $(1/2, -73)$
   (f) $(0, 4)$ and $(-7, 0)$

9. Plot the lines in Exercise 8.
10. Find the equation of the line whose y-intercept is 3 and whose x-intercept is -5.
11. Find the equation of a line whose y-intercept is 4 and such that the area of a triangle formed by the line and the two axes is 20 square units.
12. The area of a triangle formed by the line and the two axes is 40 and the slope of the line is -5. Find the equation of the line.
13. Find the length of the portion of the line $7x + 12y = 84$ that is cut off by the two axes.
14. Plot the points $(-1, -7)$, $(4, 2)$ and $(8, 4)$. Do they lie on the same line?

2.2 Linear Inequalities

In this section we would like to discuss the graphs of relationships of the forms $Ax + By \leq C$, or $Ax + By \geq C$. For simplicity, let us simply discuss inequalities of the first type; everything we say will carry over to the other. To begin, we may add (or subtract) equal quantities to both sides of an inequality without changing it. If we multiply both sides of an inequality by a positive number, the inequality will be unchanged; if multiplied by a negative number, it will be reversed. In either case, we can reduce the inequality to one of three types:

1. $x \leq a$ or $x \geq a$,
2. $y \leq b$ or $y \geq b$,
3. $y \leq mx + b$ or $y \geq mx + b$.

In the first case, $x = a$ is a vertical line. For all points to its left, $x < a$; for all points to its right, $x > a$. In the second case, $y = b$ is a horizontal line. For all points below the line, $y < b$; for all points above the line, $y > b$. In the third case, we have similar behavior. That is, if $P(x_1, y_1)$ is below the line $y = mx + b$, then P lies on some line parallel to the original line, which will have equation $y = mx + b_1$, and b_1 is less than b. Thus at $x = x_1$, $y = mx_1 + b$, on the original line, and $y_1 = mx_1 + b_1$. Since $b_1 < b$, $y_1 < mx_1 + b$. Thus, every straight line acts as a boundary dividing the plane into two regions. In one, $Ax + By < C$, and in the other, $Ax + By > C$. The graph of $Ax + By \leq C$ then consists of the line $Ax + By = C$ and all the points on one side of the line. The only problem is, How does one determine which is the proper side? Very simply, since all the points on one side of the line satisfy the inequality and all the points on the other side do not, we test any particular point and see whether or not it works. If it does, then all the points on that side of the line do! Usually the origin is a good point to try.

Example 8 Plot the graph of the inequality $3x + 4y \leq 12$.

First we plot the line $3x + 4y = 12$. The x-intercept is 4, the y-intercept is 3 (see figure 7a). Now we test to see if the origin satisfies the inequality. Substituting $x = 0$, $y = 0$, we obtain

$$3(0) + 4(0) \leq 12$$
$$0 \leq 12$$

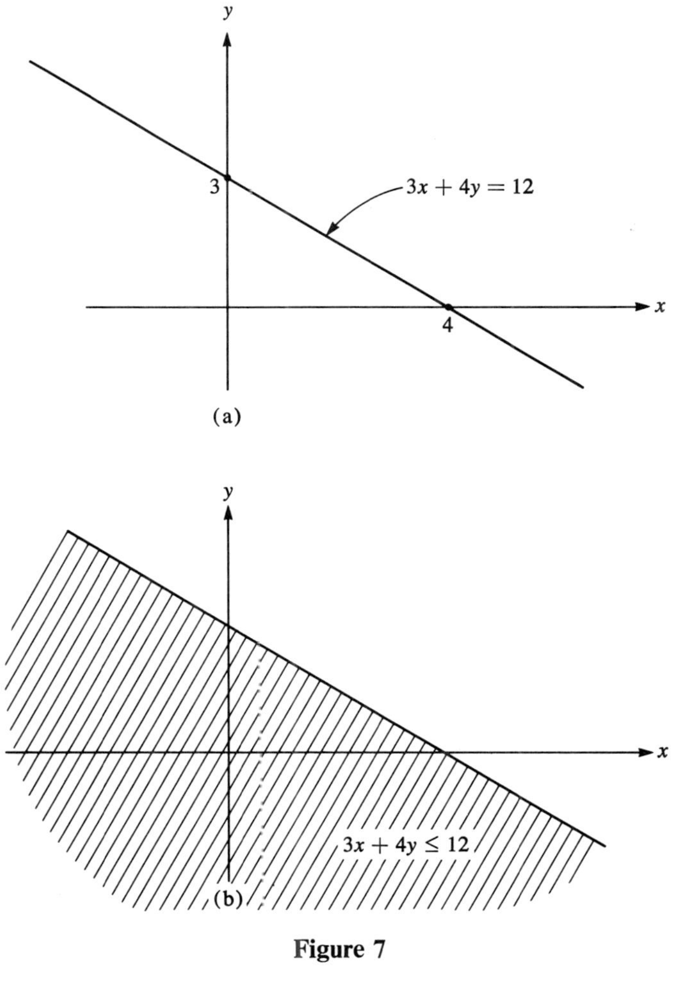

Figure 7

which is true. Hence the origin and all other points below or on the line constitute the graph shown in figure 7b. If the origin is not usable, try any other point. An example follows.

Example 9 Plot the graph of $y \leq 3x$.

The line $y = 3x$ passes through the origin. Therefore, we must try a different point. Take (0, 1); $1 \leq 0$ is not a true statement, so the inequality is satisfied by all the points below or on the line. See figure 8.

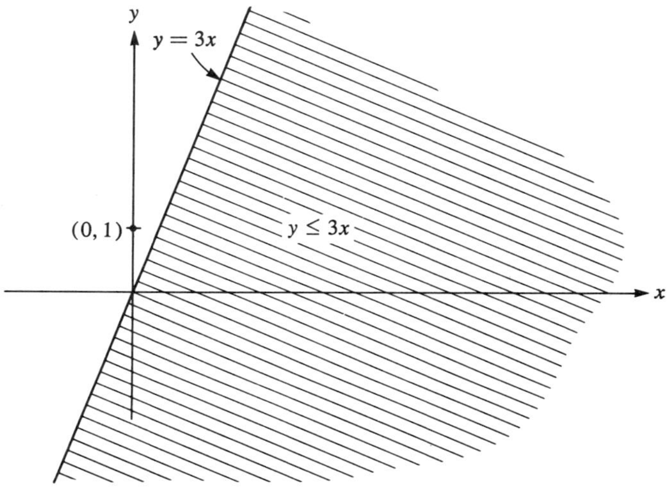

Figure 8

EXERCISES

1. Plot the graph of $2x - 5y \leq 10$.
2. Plot the graph of $3x + 11y + 8 \leq 0$.
3. Plot the set $\{(x, y) : x \geq 2\} \cap \{(x, y) : x \leq 5\}$.
4. Find the inequality which has the following region as its graph:
   (a) The boundary is a straight line of slope -3 passing through (2, 5) and the region includes the point (4, 6).
   (b) The boundary is a straight line passing through (1/2, 1/2) and $(-1, 3)$; and $(-5, 12)$ is included in the region.
   (c) All the points below or on the line $y = 7$ and to the right of or on $x = -3$.
   (d) All the points whose abscissa is between $-1/4$ and 3, inclusive, and whose ordinate is no greater than 6.

5. Plot the graph of $\{(x, y) : x \geq 1, x \leq 5 \text{ and } y \leq x\}$.
6. Plot the graph of $\{(x, y) : x \geq 1, y \geq 0 \text{ and } y \leq 5 - x\}$.

2.3 Linear Programming

One of the newest and most applicable branches of mathematics deals with the solution of problems which involve only linear equations and linear inequalities. This area is called *linear programming*. Most real problems in this discipline deal with many unknowns and involve many inequalities. We shall investigate some more or less artificial problems which can be handled without the help of computers or calculating machines but which expose the fundamental ideas which are involved.

Let us begin with the following: A company sells boxes of mixed nuts under two different brand labels. Under the regular label, a box of nuts contains 4 ounces of pecans, 8 ounces of cashews and the remainder peanuts. Under the special quality label, a box contains 1 ounce of pecans, 9 ounces of cashews and the remainder peanuts. The company can afford to use 40 ounces of pecans and 108 ounces of cashews at a time. The supply of peanuts is unlimited. Let us try to describe this situation mathematically.

$$x = \text{the number of boxes of regular to be sold}$$

$$y = \text{the number of boxes of special to be sold}$$

Now,

$$4x + y = \text{total number of ounces of pecans used}$$

$$8x + 9y = \text{total number of ounces of cashews used}$$

Not knowing the size of the box, we do not know how many ounces of peanuts will be used, but the supply of peanuts is unlimited anyway. According to the description, the number of ounces of pecans may not exceed 40 and the number of ounces of cashews may not exceed 108. Thus, we have the following set of "constraints":

$$\begin{aligned} 4x + y &\leq 40 \\ 8x + 9y &\leq 108 \\ x &\geq 0 \\ y &\geq 0 \end{aligned}$$

where the last two inequalities indicate that one cannot make a negative number of boxes. Graphically, this set of constraints restricts our attention to a set F of "feasible solutions,"

$$F = \{(x, y) : 4x + y \leq 40\} \cap \{(x, y) : 8x + 9y \leq 108\} \cap \{(x, y) : x \geq 0\} \cap \{(x, y) : y \geq 0\}$$

Geometrically, this consists of all points in the first quadrant (i.e., above the x-axis and to the right of the y-axis), which are simultaneously below the line $4x + y = 40$ and the line $8x + 9y = 108$, plus all the points on the boundaries. The graph of this region is shown in figure 9. The region is bounded by the quadrilateral $OPQR$, where O is the origin, P is the y-intercept of $8x + 9y = 108$, R is the x-intercept of $4x + y = 40$ and Q is the point of intersection of the two lines.

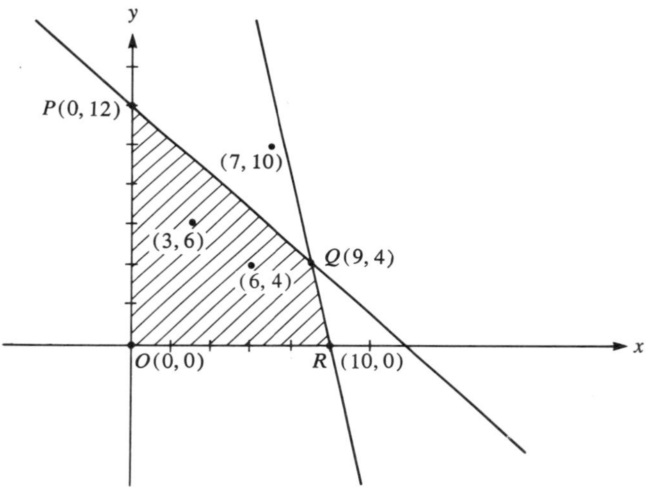

Figure 9

We can now see that the points (6, 4) and (3, 6) are feasible solutions but (7, 10) is not. It is difficult to tell from the graph whether (6, 7) is feasible or not. Checking the first inequality we have $4(6) + 7 \leq 40$, which is true, but checking the second inequality, we see $8(6) + 9(7) \leq 108$, which is not true. Hence, (6, 7) is not feasible and there is no way to make 6 boxes of regular and 7 boxes of special without running out of nuts first!

Of course, Q was obtained by solving the equations of the two lines simultaneously. That is,

$$4x + y = 40 \qquad \textbf{(1)}$$

$$8x + 9y = 108 \qquad \textbf{(2)}$$

Doubling the first equation and subtracting it from the second, we have

$$\begin{array}{r} 8x + 9y = 108 \\ -(8x + 2y = \ \ 80) \\ \hline 7y = \ \ 28 \\ \\ y = \ \ \ 4 \end{array}$$

Substituting into Eq. (1) yields

$$\begin{aligned} 4x + 4 &= 40 \\ 4x &= 36 \\ x &= 9 \end{aligned}$$

To be on the safe side, check the answer by substituting into Eq. (2):

$$8(9) + 9(4) = 108$$

which is correct.

Suppose now that the company makes a profit of 40¢ on each box of regular and 30¢ on each box of special. How many of each should be made in order to maximize the profit? Calling the profit in cents M, we have

$$M = 40x + 30y$$

For example,

$$\begin{aligned} M &= 40(6) + 30(4) = 360¢ = \$3.60 \quad \text{at } (6, 4) \\ M &= 40(3) + 30(6) = 300¢ = \$3.00 \quad \text{at } (3, 6) \end{aligned}$$

and

$$M = 40(0) + 30(12) = 360¢ = \$3.60 \quad \text{at } (0, 12)$$

Is there any way to determine the point at which M is the greatest without actually checking every point in the set F? Let us see if we can find a geometric interpretation of the expression for profit (known as the objective function).

If we rewrite the objective in the form

$$\begin{aligned} 30y &= -40x + M \\ y &= (-4/3)x + M/30 \end{aligned}$$

we see that we have a line of slope $-4/3$ and y-intercept $M/30$. To find the maximum possible value of M, we must find the line parallel to $y = (-4/3)x$, which intersects the set F and has the largest possible

y-intercept. In figure 10, several members of the family of lines are shown. For $M = 150$ and for $M = 360$, the line crosses F and so many feasible points give these values of M. For $M = 540$, the line completely misses F and so there are no feasible points for which this value can be obtained. Notice that as M increases the line moves upward. From the geometry, it appears that the maximum possible value of M will occur at the value for which we pass from F to F'. That is, in this case, the line which goes through $Q(9, 4)$ will have the greatest feasible value of M. In fact, at $(9, 4)$, $M = 40(9) + 30(4) = 480¢ = \$4.80$. Thus, the optimal production scheme is to make 9 boxes of regular, 4 boxes of special and show a profit of \$4.80.

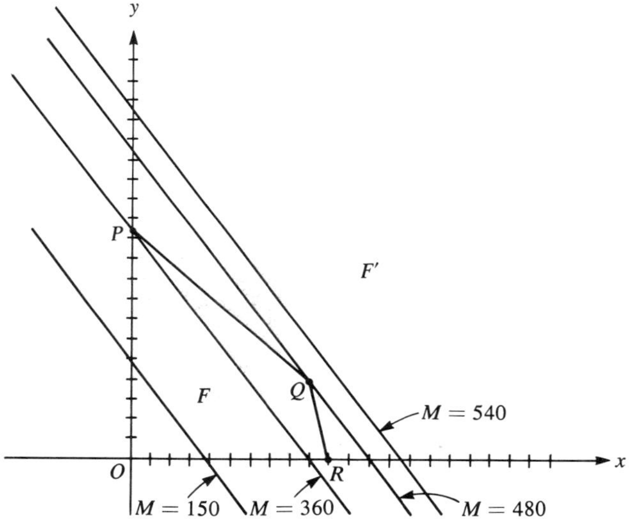

Figure 10

Let us do some further analysis of the problem. Consider table 1.

Table 1

Point	Pecans Used	Cashews Used	Pecans Left	Cashews Left	Profit
(0, 0)	0 oz	0 oz	40 oz	108 oz	0
(3, 6)	18 "	78 "	22 "	30 "	\$3.00
(6, 4)	28 "	84 "	12 "	24 "	3.60
(0, 12)	12 "	108 "	28 "	0 "	3.60
(10, 0)	40 "	80 "	0 "	28 "	4.00
(9, 4)	40 "	108 "	0 "	0 "	4.80

From the table it appears that the point of optimum profit is attained when the supply of both kinds of nuts is exhausted. However, this is *not* always the case. In fact, let us reconsider the problem assuming three different possible profit structures: Case I, the one we have already considered; Case II, the profit on regular is 5¢/box and the profit on special is 50¢/box; Case III, the profit on regular is 50¢/box and the profit on special is 5¢/box. We now have three different profit functions:

$$M_1 = 40x + 30y$$

$$M_2 = 5x + 50y$$

$$M_3 = 50x + 5y \qquad \text{(Consider table 2.)}$$

Table 2

Point	Profit Case I	Profit Case II	Profit Case III	
(0, 0)	0	0	0	
(3, 6)	$3.00	$3.15	$1.80	
(6, 4)	3.60	2.30	3.20	
(0, 12)	3.60	6.00	.60	Cashews Used Up
(10, 0)	4.00	.50	5.00	Pecans Used Up
(9, 4)	4.80	2.45	4.70	Both Used Up

Although tables 1 and 2 are in no sense exhaustive of all possibilities, they seem to indicate that the maximum in every case is attained at a point at which at least one item is used up. This, of course, is sensible since if some of each were left over, we could pack a little more and increase the profit. Of course, we are assuming that we could pack up a fractional number of boxes. If we were to restrict attention to solutions in whole numbers only, we would enter the realm of a closely related subject called *integer programming*. At any rate, the result also makes sense geometrically, since the points at which maxima are attained correspond to vertices of the region F. Clearly, a line moving upward must last touch F at one of the corners. What then decides which vertex is the right one? The answer is simply the slope of the objective. Figure 11 shows the three objectives we have considered and illustrates the reason that each is optimized at a different vertex.

One of the first problems dealt with by the technique of linear programming was called the diet problem. A simplified version of the diet problem is the following. A diet supplement is made from a mixture of wheat germ and soy flour. The composition of one ounce of each, in terms

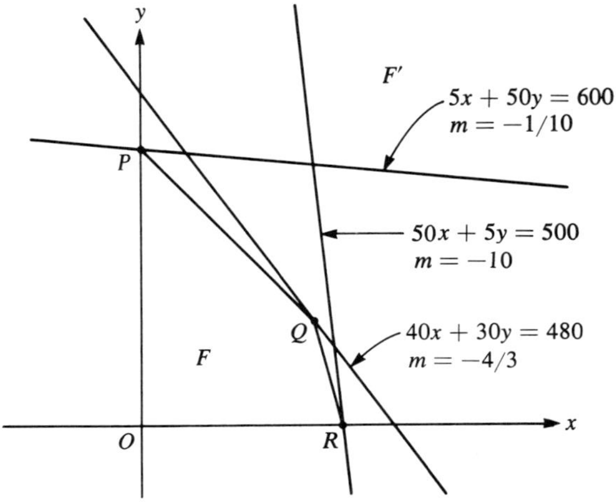

Figure 11

of Vitamin A, Vitamin B and Vitamin C, is shown in table 3. (The figures are purely imaginary.)

Table 3

	Vitamin A	Vitamin B	Vitamin C	Cost
Wheat Germ	2 units	5 units	3 units	11¢/oz
Soy Flour	1 "	3 "	2 "	6¢/oz

Find the mixture of the two ingredients satisfying the following conditions. First, the mixture must contain at least 20 units of A, 58 units of B and 36 units of C. Second, the cost should be as low as possible. To solve the problem, we shall proceed as in the previous example.

x = number of ounces of wheat germ in the mixture

y = number of ounces of soy flour in the mixture

Then we have the following constraints:

$$2x + y \geq 20 \quad \text{(units of Vitamin A)}$$
$$5x + 3y \geq 58 \quad \text{(units of Vitamin B)}$$
$$3x + 2y \geq 36 \quad \text{(units of Vitamin C)}$$
$$x \geq 0$$
$$y \geq 0$$

where the last two constraints again express the condition that one cannot mix up a negative number of ounces of a substance. The objective is to minimize the cost, $C = 11x + 6y$. Graphing the region F of feasible solutions (see figure 12), we see that the region has four vertices, (0, 20), (2, 16), (8, 6) and (12, 0). Again the optimal solution would occur at one of the vertices. Checking these four points, we obtain the following: at (0, 20), $C = 120¢ = \$1.20$; at (2, 16), $C = 118¢ = \$1.18$; at (8, 6), $C = 124¢ = \$1.24$; and at (12, 0), $C = 132¢ = \$1.32$. Thus, we claim that the most economical mixture is 2 ounces of wheat germ to every 16 ounces of soy flour with a resulting cost of $1.18.

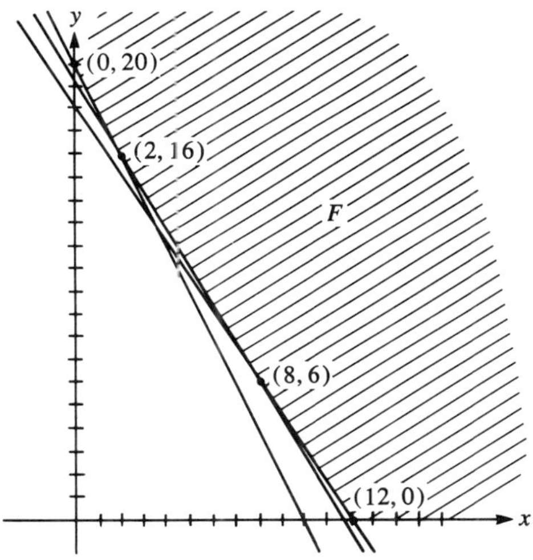

Figure 12

It is interesting to note that if the price of wheat germ were to drop to 10¢/ounce, we would be faced with an odd situation. In that case, $C = 10x + 6y$. Checking the vertices we find the following: at (0, 20), $C = \$1.20$; at (2, 16), $C = \$1.16$: at (8, 6), $C = \$1.16$; and at (12, 0), $C = \$1.20$. The optimal point then occurs at two adjacent vertices. Therefore, either point is a suitable choice. Moreover, any point on the line segment joining them is also an optimal point! This happens because the objective is parallel to the segment and hence every point on the segment gives the same y-intercept and hence the same value to C. The equation of the line in question is $10x + 6y = 116$, which passes through (2, 16) and (8, 6). In addition, the point halfway between is (5, 11) and this also gives the value $1.16 to the objective as will any other point on the segment.

EXERCISES

1. Plot the regions described by the following sets of inequalities:

   (a) $x \geq 1$
   $x \leq 5$
   $y \geq 2$
   $y \leq 2x + 6$

   (b) $x \leq 5$
   $y \geq -2$
   $y \leq x + 6$

   (c) $x + 3y \geq 4$
   $4x + y \geq 10$
   $x \geq 0$
   $y \geq 0$

   (d) $2x + y \leq 10$
   $2x + y \geq 5$
   $y \leq 3x$
   $y \geq (1/2)x$

   (e) $x + 4y \leq 40$
   $x + 2y \leq 30$
   $x \leq 23$
   $x \geq 0$
   $y \geq 0$

2. Find a set of inequalities whose solution set is a quadrilateral with vertices (1, 6), (3, 7), (1/2, 2) and (5, 0).
3. Find a set of inequalities whose solution set is a pentagon with vertices (0, 0), (0, 10), (5, 9), (6, 6) and (7, 0).
4. For Exercise 1, (a)–(e), find the maximum and minimum values of $P = 3x + 5y$ and $C = 2x - 7y$ on each of the regions.
5. Solve the mixed-nut example in the text if each box must weigh one pound (16 ounces) and if there are only 66 ounces of peanuts available.
6. A candy manufacturer makes two basic types of merchandise. The pertinent information is displayed in the following table:

Type	Coconut Used/Box	Chocolate Used/Box	Profit/Box
Coconut-covered Chocolate	1 ounce	14 ounces	\$2.00
Chocolate-covered Coconut	10 "	1 "	3.00

He has available 150 ounces of coconut and 154 ounces of chocolate. How many boxes of each should he make in order to maximize his profit?

7. After Christmas, a toy store has a collection of leftover ornaments to get rid of. The store owner decides to make up two types of "bargain packs" and run a sale. The first type of pack contains 2 strings of lights and 5 angels and sells for \$1. The second type of pack contains 3 strings of lights and 3 angels and sells for \$2. He has 42 strings of lights and 78 angels available. How many of each type should he make up in order to maximize his receipts? If he wishes to maximize his receipts at a time when he has used up all the ornaments, what possible prices could he charge for the second type of pack? (The price of the first is fixed at \$1.)

2.4 Enumeration of Vertices

From the examples in the preceding section the form of a general linear programming problem in two variables seems clear. The problem always consists of a statement of the form "maximize (or minimize) some objective, subject to certain constraints." To be more precise, the statement is usually presented in the following format:

$$\text{Maximize (or minimize): } Q = Dx + Ey$$

$$\text{Subject to: } A_1x + B_1y \leq C_1 \qquad (1)$$

$$A_2x + B_2y \leq C_2 \qquad (2)$$

$$\vdots$$

$$A_kx + B_ky \leq C_k \qquad (k)$$

although some of the inequalities may appear to be reversed.

The method of attack also seems to be clear. That is, take all possible pairs of constraint inequalities, replace them by equations and solve simultaneously, thereby obtaining all points of intersection. By the geometry of the situation, some of these intersection points will be the vertices of F and the others will lie outside the region of feasible solutions. So one must then check each point of intersection and ascertain whether or not it lies in F. This is accomplished by simply substituting the coordinates in each of the k constraint inequalities. Those which satisfy them all are of interest, the others are not. For the points of interest, we evaluate Q. The largest number obtained is the maximum possible value of Q and the smallest is the minimum possible value of Q (subject to the constraints, of course). This method of "enumeration of vertices" is perhaps best illustrated by a simple example. Consider the following problem:

$$\begin{aligned} \text{Maximize: } Q &= 3x - 4y \\ \text{Subject to: } x + 4y &\leq 10 \quad &(1) \\ 4x + y &\leq 10 \quad &(2) \\ x &\geq 0 \quad &(3) \\ y &\geq 0 \quad &(4) \end{aligned}$$

We see that we shall have to solve six systems of two equations in two unknowns. In particular they are Eqs. (1) and (2), Eqs. (1) and (3), Eqs. (1) and (4), Eqs. (2) and (3), Eqs. (2) and (4) and Eqs. (3) and (4). Solving Eqs. (1) and (2) simultaneously yields

$$\begin{aligned} x + 4y &= 10 \\ 4x + y &= 10 \end{aligned}$$

The reader may check that the point of intersection is (2, 2) Solving Eqs. (1) and (3) yields

$$\begin{aligned} x + 4y &= 10 \\ x &= 0 \end{aligned}$$

It is easy to see that the point of intersection is (0, 2½). Continuing in this manner, we may summarize our results in tabular form (see table 4).

Table 4

Equation Pair	Point	Is the Point in F?	Value of Q	Remarks
(1) and (2)	(2, 2)	Yes	-2	—
(1) and (3)	(0, 2½)	Yes	-10	Minimum value
(1) and (4)	(10, 0)	No	—	Does not satisfy Eq. (2)
(2) and (3)	(0, 10)	No	—	" " " " (1)
(2) and (4)	(2½, 0)	Yes	7½	Maximum value
(3) and (4)	(0, 0)	Yes	0	—

We have additionally noted in the last section that when the maximum of the objective function occurs at two different vertices, it is the same at every point on the line segment joining them. In the rest of this section, we shall indicate how one can prove that this method does in fact work and that the existence of two solutions implies a whole continuum of solutions. To do so, we shall introduce some terminology which we will also find useful in the study of probability.

Definition 1 Let $x_1, x_2, \ldots, x_k$ be any k numbers. Let $a_1, a_2, \ldots, a_k$ be k numbers whose sum is 1 and none of which is negative. The sum $a_1x_1 + a_2x_2 + \cdots + a_kx_k$ is called the weighted average of the x's, and the a's are called the weights.

For example, the weighted average of 6, 8 and -3 with weights 1/2, 1/3 and 1/6 is

$$(1/2)6 + (1/3)8 + (1/6)(-3) = 3 + 8/3 - 1/2 = 31/6$$

Weighted averages are commonly used in business. For example, a salesman made 100 sales in a single day. Fifty sales were ties selling for \$5 each, 30 sales were shirts selling for \$8 each and 20 sales were hats selling for \$15 each. What was the average sale? His total tie sale was (50)(5), his total shirt sale was (30)(8) and his total hat sale was (20)(15). Adding these and dividing the sum by 100, the total number of sales, we get

$$[(50)(5) + (30)(8) + (20)(15)]/100 = (.5)(5) + (.3)(8) + (.2)(15)$$

which is a weighted average of 5, 8 and 15 with the weights 0.5, 0.3 and 0.2. The average sale is \$7.90. If one were to simply average 5, 8 and 15, that is $(5 + 8 + 15)/3$, the result is \$9.33. This is, of course, not as good an average as the weighted average, since the weighted average reflects the predominance of small sales. The simple average is just a special case of a weighted average but with equal weights. That is, in this case dividing the total by three is the same as multiplying each number by 1/3 and adding the result. Or, in general, adding k numbers and dividing the sum by k amounts to taking a weighted average with all the weights equal to $1/k$.

Definition 2 The symbol $a(x, y)$, read "a times (x, y)," is defined to be the point with coordinates (ax, ay).

Definition 3 Given two points (x_1, y_1) and (x_2, y_2), we define $(x_1, y_1) + (x_2, y_2) = (x_1 + x_2, y_1 + y_2)$.

Definition 4 A weighted average of k points in the plane is $a_1(x_1, y_1) + a_2(x_2, y_2) + \cdots + a_k(x_k, y_k) = (a_1x_1 + \cdots + a_kx_k, a_1y_1 + \cdots + a_ky_k)$. That is, a point is the weighted average of k other points, if each coordinate of the point is the weighted average of the separate coordinates (with the same weights). For example,

$$(3, -2) = (1/2)(4, -2) + (1/3)(6, 3) + (1/6)(-6, -12)$$

We shall now give some theorems about weighted averages of points, which may shed some light on the geometric meaning of the term.

THEOREM 2 If P_1 and P_2 are two points, then any point on the line segment P_1P_2 is a weighted average of the end points, and conversely.

Proof Let $P_1 = (x_1, y_1)$ and $P_2 = (x_2, y_2)$. Let $P(x, y)$ be any point on the segment P_1P_2. Let the ratio $P_1P : P_1P_2 = f$; now the ratio $PP_2 : P_1P_2 = g$ and $f + g = 1$. Referring to figure 13, we see that triangles P_1RP and P_1SP_2 are similar. Hence, $P_1R : P_1S = P_1P : P_1P_2 = RP : SP_2 = f$. $P_1R = x - x_1$, $P_1S = x_2 - x_1$, $RP = y - y_1$, and $SP_2 = y_2 - y_1$. Therefore,

$$\frac{x - x_1}{x_2 - x_1} = f \quad \text{and} \quad \frac{y - y_1}{y_2 - y_1} = f$$

$$x - x_1 = fx_2 - fx_1 \quad \text{and} \quad y - y_1 = fy_2 - fy_1$$

$$x = fx_2 + (1 - f)x_1 \quad \text{and} \quad y = fy_2 + (1 - f)y_1$$

$$x = gx_1 + fx_2 \quad \text{and} \quad y = gy_1 + fy_2$$

so that

$$(x, y) = g(x_1, y_1) + f(x_2, y_2)$$

where f, g are nonnegative and $f + g = 1$. As a special case of this theorem, we have the following: The midpoint of a line segment is the average

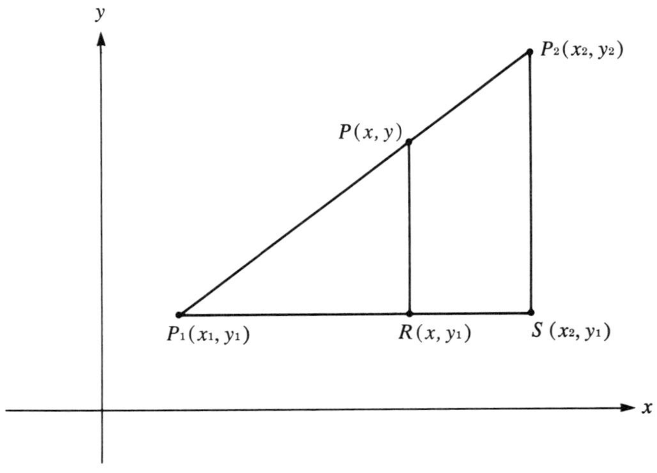

Figure 13

of the end points. So (3, 5) is the point halfway between (5, 2) and (1, 8).

As an immediate consequence of Theorem 2, we have the following corollary.

Corollary 1 If $Q = Dx + Ey$ and at P_1, $Q = Q_1$ and at P_2, $Q = Q_2$, then at $fP_1 + gP_2$, $Q = fQ_1 + gQ_2$.

Proof $Q_1 = Dx_1 + Ey_1$, $Q_2 = Dx_2 + Ey_2$. Let $P = fP_1 + gP_2 = (fx_1 + gx_2, fy_1 + gy_2)$. At P,

$$\begin{aligned} Q &= D(fx_1 + gx_2) + E(fy_1 + gy_2) = Dfx_1 + Dgx_2 + Efy_1 + Egy_2 \\ &= f(Dx_1 + Ey_1) + g(Dx_2 + Ey_2) = fQ_1 + gQ_2 \end{aligned}$$

Corollary 2 If the objective in a linear programming problem has the same value at two vertices, then it has the same value along an entire segment.

Proof Let $Q = Dx + Ey$ be equal to m at the points $P_1(x_1, y_1)$ and $P_2(x_2, y_2)$. Then at any point on P_1P_2, $Q = fm + gm = (f + g)m = m$, since $f + g = 1$.

Notice that Corollary 2 is just a hard way of saying, "The weighted average of a collection of equal numbers is simply the number, regardless of the weights." Suppose we take the weighted average of a collection of unequal numbers—is there anything we can say about the possible result? The following lemma gives us some of the answer.

Lemma 1 Let $x_1, x_2, \ldots, x_k$ be k numbers and for simplicity assume that x_1 is the smallest and x_k is the largest. Then any weighted average of the x's is no less than x_1 and no more than x_k.

Proof Let $w = a_1x_1 + a_2x_2 + \cdots + a_kx_k$, the weighted average. If in this expression all the x's are replaced by the largest x, the result must be at least as large. That is,

$$\begin{aligned} w &\leq a_1x_k + a_2x_k + \cdots + a_kx_k \\ &= (a_1 + a_2 + \cdots + a_k)x_k = x_k \end{aligned}$$

Similarly, replacing all the x's by x_1 shows w at least as large as the smallest x. Thus, $x_1 \leq w \leq x_k$ as claimed.

Returning now to the geometry, we have shown that the weighted average of two points must be a point on the line segment joining them.

We shall now show that the weighted average of three points is a point in the triangle determined by the points.

THEOREM 3 Let $P_1(x_1, y_1)$, $P_2(x_2, y_2)$ and $P_3(x_3, y_3)$ be any three non-collinear points. Then every point in the interior of the triangle $P_1P_2P_3$ can be expressed in the form $f_1P_1 + f_2P_2 + f_3P_3$, where $f_1 + f_2 + f_3 = 1$ and $f_j \geq 0$, $j = 1, 2, 3$.

Proof Let $P(x, y)$ be any point in the interior of the triangle. Draw the line segment from P_1 through P and extend it until it intersects P_2P_3. Call the point of intersection R (see figure 14). Now $P = aP_1 + bR$, where $a, b \geq 0$ and $a + b = 1$, and $R = cP_2 + dP_3$, where $c, d \geq 0$ and $c + d = 1$, by Theorem 2. Thus,

$$P = aP_1 + b(cP_2 + dP_3) = aP_1 + bcP_2 + bdP_3$$

where obviously a, bc, $bd \geq 0$ and $a + bc + bd = a + b(c + d) = a + b = 1$.

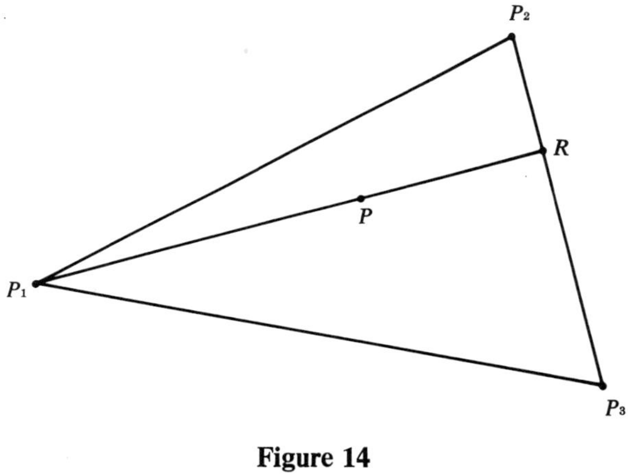

Figure 14

It now follows from Lemma 1 that the largest and smallest values of a linear objective in any triangle are attained on the vertices. This is stated formally in the next corollary.

Corollary 3 Let $Q_j = Ax_j + By_j$, $j = 1, 2, 3$; then if $P(x, y)$ is any point in the triangle $P_1\,P_2\,P_3$, there exist a, b, $c \geq 0$, $a + b + c = 1$, and such that $Ax + By = aQ_1 + bQ_2 + cQ_3$; Hence $Ax + By$ is less than or equal to the largest of the three numbers Q_1, Q_2, Q_3 and greater than or equal to the smallest.

This result now extends easily to all polygons. However, to avoid complications in the proof, let us simply restrict attention to convex polygons. First, we make the concept precise.

Definition 5 A region is called *convex* if given any two points P_1 and P_2 in the region, every point on the segment P_1P_2 is also in the region. (See figure 15.) We now complete this development.

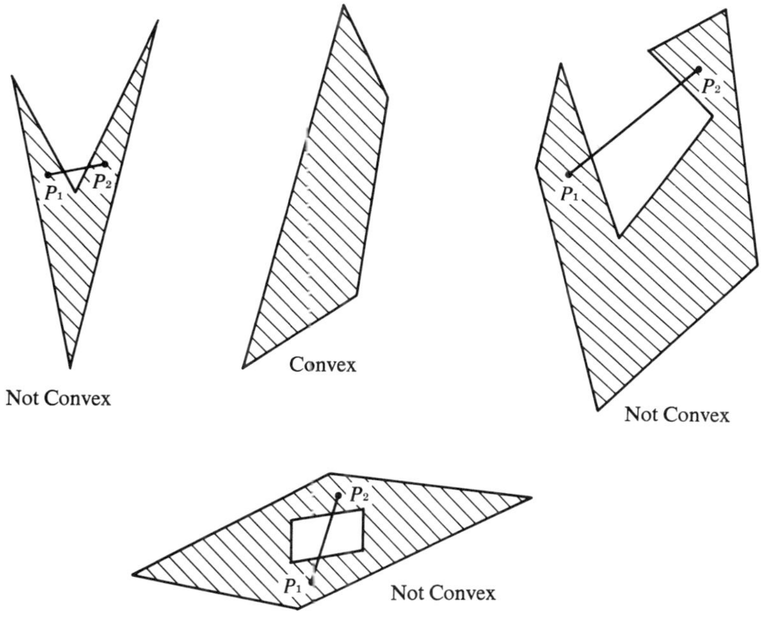

Figure 15

THEOREM 4 Let $P_1P_2 \ldots P_k$ be any bounded convex polygon. Then any point $P(x, y)$ in the polygon is a weighted average of the vertices, and consequently the maximum and minimum values of the objective $Ax + By$ are attained on vertices of the polygon.

Proof Draw lines $P_1P_3, P_1P_4, \ldots, P_1P_{k-1}$ (see figure 16). Now every point in the polygon lies in the interior of one of the triangles formed and is thus a weighted average of the vertices, all of which are vertices of the original polygon. The result is then an immediate consequence of Theorem 3 and Corollary 3.

We know that the boundary of a set of feasible solutions to a linear programming problem is composed of line segments. Hence, if the set is bounded, it is a polygon. What we need to show is that it is a convex

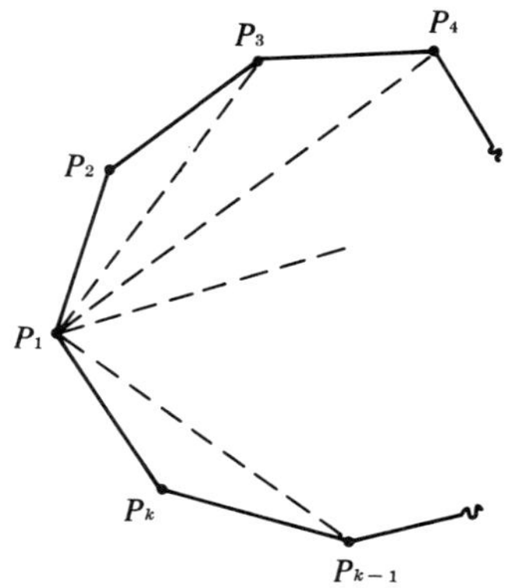

Figure 16

polygon. Incidentally, if the set of feasible solutions is unbounded, it is possible to show that the extreme values either occur on the boundary or do not exist. It is also not necessary that the polygon be convex, since any polygon can be divided into triangles, and hence Theorem 4 does not really depend upon convexity. Nevertheless, the set of feasible solutions is convex and is easy to see.

THEOREM 5 The set F of feasible solutions to the linear programming problem is convex.

Proof The set F consists of points (x, y) for which $A_jx + B_jy \leq C_j$, $j = 1, 2, 3, \ldots, k$. Let $P_1(x_1, y_1)$ and $P_2(x_2, y_2)$ be points in F. Then $A_jx_1 + B_jy_1 = L_{1j} \leq C_j$ and $A_jx_2 + B_jy_2 = L_{2j} \leq C_j$. Now if (x, y) is any point on the segment P_1P_2, $A_jx + B_jy$ will be a number between L_{1j} and L_{2j} and hence less than C_j. Since this is true for all j, $(x, y) \in F$ and the proof is completed.

EXERCISES

1. Maximize the objective $w = 3x - y$, subject to

$$
\begin{aligned}
x + 2y &\geq 4 \\
x + y &\leq 12 \\
2x + 4y &\leq 30 \\
x &\geq 0 \\
y &\geq 0
\end{aligned}
$$

2. A tailor has 80 square yards of cotton material and 120 square yards of wool material. A suit requires 1 sq yd of cotton and 3 sq yd of wool. A dress requires 2 sq yd of each. How many of each garment should the tailor make in order to maximize his income, if a suit sells for \$80 and a dress for \$50?

3. In an ecology drive, the town of Cleanup sets a goal of processing at least 40 tons of paper, 30 tons of aluminum cans and 50 tons of glass bottles each month. Roundups of garbage made in residential and commercial districts show the following distribution of material:

Type of District	Paper	Aluminum	Glass	Cost/Pickup
Commercial	3 tons	1 ton	1 ton	\$50
Residential	1 ton	1 "	5 "	25

How many pickups of each kind should be made in order to minimize the cost of the project?

4. A hospital patient is being fed intravenously. His diet is a mixture of two fluids, A and B. One unit of A contains 5 grams of protein and 100 calories, and one unit of B contains 10 grams of protein and 50 calories. His total protein intake must be at least 50 grams and may not exceed 100 grams. In addition, for proper metabolism, more of fluid B than fluid A must be administered. Find the mixture of the two fluids that maximizes his caloric intake.

5. The candy manufacturer in Exercise 6, Section 2.3, decides to add 6 ounces of peanuts to each type of candy. Resolve the problem assuming he wishes to use only 180 ounces of peanuts.

6. Plot the convex polygon $ABCDEF$, with vertices $A(2, 0)$, $B(5, 5)$, $C(1, 10)$, $D(-4, 10)$, $E(-5, 5)$ and $F(-2, -1)$. Divide the polygon into triangles as described in the proof of Theorem 3.

7. Plot the nonconvex polygon $ABCDEFGH$, with vertices $A(2, 0)$, $B(5, 5)$, $C(1, 10)$, $D(2, 4)$, $E(-4, 10)$, $F(-5, 5)$, $G(-2, -1)$, and $H(-1/2, 3)$. Show how this polygon may be divided into triangles without introducing any new vertices.

8. Given a triangle ABC, $A(-1, 0)$, $B(5, 8)$ and $C(7, -4)$, find the equations of the lines which are the extensions of the medians of the triangle. Show that $M = (1/3)A + (1/3)B + (1/3)C$ is the point of intersection of the three medians.

9. Express (3, 12/5) as a weighted average of (2, 4) and (7, −4). HINT: Write the point in the form $f(2, 4) + (1 - f)(7, -4)$.

10. Show that (3, 2) is not on the segment joining (2, 4) and (7, −4). HINT: Proceed as in Exercise 9. Show that you have two equations in one unknown and the solutions are not the same.

3

Systems of Equations

3.1 Simple Systems

At the end of the last chapter, we showed that any point in the interior of a convex polygon could be expressed as a weighted average of the vertices. However, we did not really describe how the proper weights could be found. For example, (5, 3) is in the interior of the triangle with vertices (1, 1), (8, 2), (4, 12). Therefore, numbers f, g, h exist with the property that

$$f(1, 1) + g(8, 2) + h(4, 12) = (5, 3)$$

and

$$f + g + h = 1$$

This means that

$$\begin{aligned} f + 8g + 4h &= 5 \\ f + 2g + 12h &= 3 \qquad \text{(System 1)} \\ f + g + h &= 1 \end{aligned}$$

This gives us a system of three equations in three unknowns. It is assumed that the reader has some familiarity with systems of this type and if pressed could actually solve it. However, later in this chapter we shall develop the most efficient technique for solving such systems. We shall also consider some less usual cases. For example, express (4, 7) as a weighted average of (2, 4) and (8, 13). This leads to $f(2, 4) + g(8, 13) = (4, 7)$ and $f + g = 1$, which produces

$$\begin{aligned} 2f + 8g &= 4 \\ 4f + 13g &= 7 \qquad \text{(System 2)} \\ f + g &= 1 \end{aligned}$$

This is a system with more equations than unknowns and may be less familiar. As a third example, consider the problem of finding (5, 9) as a weighted average of (1, 1), (6, 2), (6, 14) and (−1, 8). Geometrically, the thing to do is to sketch the quadrilateral with the four points as vertices, divide it into two triangles, decide which triangle contains (2, 5) and write the point as a weighted average of the vertices of this triangle (weight the fourth vertex zero). However, this is not the only solution. Algebraically, all that is required is

$$f(1, 1) + g(6, 2) + h(6, 14) + k(-1, 8) = (5, 9)$$

and

$$f + g + h + k = 1$$

This yields the following system:

$$\begin{aligned} f + 6g + 6h - k &= 5 \\ f + 2g + 14h + 8k &= 9 \qquad \text{(System 3)} \\ f + g + h + k &= 1 \end{aligned}$$

Here, we have fewer equations than unknowns.

In this chapter, we shall consider the general problem of solving m equations in n unknowns in a systematic manner. We shall only deal with systems of "linear" equations, that is, equations in which the unknowns appear only to the first power. If the equations involve only two unknowns, then we know that graphically they represent straight lines and we are looking for the point of intersection of the lines. In this case, three things can occur. First, the lines could have one point of intersection (even if there are several lines). Algebraically, this would mean that the system of equations has a unique solution. In other words, there exists exactly one ordered pair which satisfies all the equations simultaneously. A second possibility is that the lines have no point in common. When one has three lines, for example, it is usual to have them form a triangle. That is, each intersects the other but there is no point common to all three. In this case, we say the equations are "incompatible" and there is no solution. Even two equations in two unknowns can be incompatible if they represent parallel lines, since parallel lines never intersect. The third possibility is that all the equations are different representations of the same line. In that case any point on the line is a solution of the system and hence we have an infinite number of solutions. Such a system is called "redundant." When more than two unknowns are present, many variations on this theme are possible; however, all reduce to one of these three situations: The system has a unique solution; the system is incompatible; or the system is redundant.

Let us then look at the problem of solving the equations. The underlying idea is to replace the system of equations with an equivalent system in which the roots can be identified. By equivalent systems we mean two systems having identical roots. For example, solve

$$x + y = 3$$
$$x - y = 4$$

Adding the two equations, we have $2x = 7$, $x = 3\frac{1}{2}$, which we substitute into the first equation, yielding $3\frac{1}{2} + y = 3$, $y = -1/2$. Now, what have we really done? We have added the two equations, producing a new equation, $2x = 7$. We have combined this with the first equation to form a new system:

$$x + y = 3$$
$$2x \qquad = 7$$

This new system is equivalent to the original one and we can obtain its roots by inspection (almost).

The important question is, What operations can you perform on a system of linear equations that will produce a new system having the same roots? There are three allowable operations:

1. Any two equations may be interchanged.
2. Any equation may be multiplied (or divided) by any number other than zero.
3. Any equation may be replaced by the equation obtained by adding to it any multiple of any other equation.

A few parenthetical remarks are appropriate here. First, in (1), it seems obvious that the order in which you write the equations should be immaterial. Second, in (2), you cannot multiply by zero, since this would eliminate the equation. Third, in (3), adding any multiple also allows subtracting any multiple, since this amounts to simply adding a negative multiple. This operation is the most important one as you will see in the first example.

In the following, we have labeled the equations for easy identification:

$$x + y + z = 2 \qquad \textbf{(1)}$$
$$x + 2y - z = 6 \qquad \textbf{(2)}$$
$$x - y + 2z = -3 \qquad \textbf{(3)}$$

First, we replace Eq. (2) by Eq. (2) − Eq. (1) and we replace Eq. (3) by Eq. (3) − Eq. (1). By performing these operations, we eliminate x from all but the first equation.

$$x + y + z = 2 \qquad \textbf{(1)}$$
$$y - 2z = 4 \qquad \textbf{(2)}$$
$$-2y + z = -5 \qquad \textbf{(3)}$$

Next, we replace Eq. (3) by Eq. (3) + 2 times Eq. (2). This has the effect of giving us a "triangular configuration" with the last equation having only one unknown; the next to last, two unknowns; and the first, all three.

$$x + y + z = 2 \qquad \textbf{(1)}$$
$$y - 2z = 4 \qquad \textbf{(2)}$$
$$-3z = 3 \qquad \textbf{(3)}$$

We may now read off the solutions by "back substitution." Solving Eq. (3) yields $z = -1$. Substituting this into Eq. (2), we have $y + 2 = 4$,

$y = 2$. Substituting these two values into Eq. (1) yields $x + 2 - 1 = 2$, $x = 1$. Thus, $x = 1$, $y = 2$ and $z = -1$.

Let us recapitulate what has been done, for this is the method to be used. First, we added suitable multiples of the first equation to each of the others in order to eliminate the first unknown from all but the first equation. Next, we added suitable multiples of the second equation to all those below it, in order to eliminate the second unknown from all of them. In this case, of course there is only one equation below the second. If there were more unknowns, we would continue in this manner, aiming for the triangular form. For example,

$$\begin{aligned} x - y + z + 2w &= 0 && \textbf{(1)}\\ x + y + 2z - 3w &= 13 && \textbf{(2)}\\ x \qquad - z + w &= -4 && \textbf{(3)}\\ x - y + 2z \qquad &= 7 && \textbf{(4)} \end{aligned}$$

Clearing x from Eqs. (2), (3) and (4), we have

$$\begin{aligned} x - y + z + 2w &= 0 && && \textbf{(1)}\\ 2y + z - 5w &= 13 && \text{[Obtained from Eq. (2) − Eq. (1)]} && \textbf{(2)}\\ y - 2z - w &= -4 && \text{[Obtained from Eq. (3) − Eq. (1)]} && \textbf{(3)}\\ z - 2w &= 7 && \text{[Obtained from Eq. (4) − Eq. (1)]} && \textbf{(4)} \end{aligned}$$

Clearing y from Eqs. (3) and (4) yields

$$\begin{aligned} x - y + z + 2w &= 0\\ 2y + z - 5w &= 13\\ -\frac{5}{2}z + \frac{3}{2}w &= \frac{-21}{2} && \text{[Obtained from Eq. (3) − 1/2 Eq. (2)]}\\ z - 2w &= 7 \end{aligned}$$

Multiplying Eq. (3) by -2 (in order to eliminate fractions), we obtain

$$\begin{aligned} x - y + z + 2w &= 0 && \textbf{(1)}\\ 2y + z - 5w &= 13 && \textbf{(2)}\\ 5z - 3w &= 21 && \textbf{(3)}\\ z - 2w &= 7 && \textbf{(4)} \end{aligned}$$

Clearing z from Eq. (4) yields

$$\begin{aligned} x - y + z + 2w &= 0 & \quad \textbf{(1)} \\ 2y + z - 5w &= 13 & \quad \textbf{(2)} \\ 5z - 3w &= 21 & \quad \textbf{(3)} \\ -\frac{7}{5}w &= \frac{14}{5} \quad \text{[Obtained from Eq. (4) } - 1/5 \text{ Eq. (3)]} & \quad \textbf{(4)} \end{aligned}$$

We now have the triangular pattern and upon back substitution,

$$w = -2 \qquad 5z + 6 = 21 \qquad z = 3 \qquad 2y + 3 + 10 = 13$$
$$y = 0 \qquad x - 0 + 3 - 4 = 0 \qquad x = 1$$

It is left to the reader to substitute these numbers into the original system and see that we do indeed have the roots of the system.

EXERCISES

Solve the systems of equations in Exercises 1–6.

1. $$\begin{aligned} x + 3y &= 6 \\ x - 3y &= 12 \end{aligned}$$

2. $$\begin{aligned} x - 2y + 2z &= -3 \\ x + y - z &= 0 \\ x - 3y - 3z &= -16 \end{aligned}$$

3. $$\begin{aligned} x + 2y - 2z &= 4 \\ 2x - y + 3z &= 1 \\ 3x + z &= -2 \end{aligned}$$

4. $$\begin{aligned} 3x - 4y + 3z &= -6 \\ x - 2y + 6z &= -4 \\ 2y + 9z &= -2 \end{aligned}$$

5. $$\begin{aligned} 2x + 6y - z + 2w &= 4 \\ x - y + z - 2w &= -5 \\ 2x + y - 3z + 6w &= 3 \\ x - y + 4z &= 9 \end{aligned}$$

6. $$\begin{aligned} x - z &= 2 \\ y + w &= 9 \\ x + w &= 8 \\ x - y &= -1 \end{aligned}$$

7. Solve System 1.
8. Express (4, 3) as a weighted average of (1, 1), (7, −2) and (6, 6).

3.2 Systems with Nonunique Solutions

Suppose one is faced with the problem of solving a system in which the number of unknowns and the number of equations are not the same. We will still apply the same technique. For example,

$$x - y + z + w = 6 \qquad \textbf{(1)}$$
$$2x - 3y + 3z + w = 7 \qquad \textbf{(2)}$$
$$x + y - z - 3w = 4 \qquad \textbf{(3)}$$

In this system of three equations in four unknowns, we begin by eliminating x from the last two equations:

$$\begin{aligned} x - y + z + w &= 6 & & \\ -y + z - w &= -5 & & [\text{Eq. (2)} - 2 \text{ times Eq. (1)}] \\ 2y - 2z - 4w &= -2 & & [\text{Eq. (3)} - \text{Eq. (1)}] \end{aligned}$$

Now using Eq. (2) to eliminate y, we have

$$\begin{aligned} x - y + z + w &= 6 & & \\ -y + z - w &= -5 & & \\ -6w &= -12 & & [\text{Eq. (3)} + 2 \text{ times Eq. (2)}] \end{aligned}$$

From Eq. (3), we have $w = 2$. Substituting into Eq. (1) and Eq. (2), we have

$$\begin{aligned} x - y + z + 2 &= 6 \\ -y + z - 2 &= -5 \end{aligned}$$

yielding

$$\begin{aligned} x - y + z &= 4 \\ -y + z &= -3 \end{aligned}$$

Here we see that we can choose any value for z and still find correct values for x and y. That is, let $z = k$ and substitute into Eq. (2):

$$\begin{aligned} -y + k &= -3 \\ -y &= -3 - k \\ y &= 3 + k \end{aligned}$$

Now, substitute into Eq. (1):

$$x - (3 + k) + k = 4$$
$$x - 3 - k + k = 4$$
$$x - 3 = 4$$

Thus, the general solution to this system is $x = 7$, $y = 3 + k$, $z = k$ and $w = 2$, where k is any number whatsoever! Thus, we have a redundant system. Some possible solutions are $x = 7, y = 3, z = 0, w = 2$ (the case $k = 0$); $x = 7$, $y = 113$, $z = 110$, $w = 2$ (the case $k = 110$); $x = 7$, $y = 0, z = -3$, $w = 2$ (the case $k = -3$); and $x = 7$, $y = 3\frac{1}{4}, z = 1/4$, $w = 2$ (the case $k = 1/4$). The reader should check that all the above, including the general formula, are solutions by seeing that they do satisfy the equations. A remark is in order here about redundant systems. Usually the solution involves an arbitrary number (called a *parameter*), k, and all the values change as k changes. This system is only partly redundant in that x and w are completely determined and only y and z depend upon the parameter.

As a second example, let us consider the following:

$$x - y + z = 6 \quad \textbf{(1)}$$
$$x + 3y - 2z = 8 \quad \textbf{(2)}$$
$$4y + 4z = 11 \quad \textbf{(3)}$$
$$x - 7y + 9z = 16 \quad \textbf{(4)}$$

We begin as usual:

$$x - y + z = 6$$
$$4y - 3z = 2 \qquad \text{[Eq. (2) – Eq. (1)]}$$
$$4y + 4z = 11$$
$$-6y + 8z = 10 \qquad \text{[Eq. (4) – Eq. (1)]}$$

Eliminating y yields

$$x - y + z = 6$$
$$4y - 3z = 2$$
$$7z = 9 \qquad \text{[Eq. (3) – Eq. (2)]}$$
$$\frac{7}{2}z = 13 \qquad \text{[Eq. (4) + 3/2 times Eq. (2)]}$$

We now use the third equation to eliminate z from all the equations below it:

$$\begin{aligned} x - y + z &= 6 \\ 4y - 3z &= 2 \\ 7z &= 9 \\ 0 &= 17/2 \qquad \text{[Eq. (4) } - \text{ 1/2 times Eq. (3)]} \end{aligned}$$

The last equation is now clearly an impossible or absurd statement. No matter what values of x, y and z are chosen, $0 \neq 17/2$. But this system is equivalent to the original one. Hence, the original system is incompatible. This is *not* the inevitable result of having more equations than unknowns. In fact, if Eq. (4) of the original system had been

$$x - 7y + 9z = 15/2$$

when the triangularization had been completed the final system would have turned out:

$$\begin{aligned} x - y + z &= 6 \\ 4y - 3z &= 2 \\ 7z &= 9 \\ 0 &= 0 \end{aligned}$$

The last line is perfectly valid and we would be able to solve by back substitution in the usual manner, obtaining $x = 173/28$, $y = 41/28$ and $z = 9/7$.

EXERCISES

1. Solve System 2, p. 40.
2. Solve System 3, p. 40.
3. Solve:
$$\begin{aligned} x - y + z &= 3 \\ 2x + 2y - 3z &= 7 \\ x - 2y + z &= 8 \\ x + 2y + 3z &= 11 \end{aligned}$$
4. Solve:
$$\begin{aligned} x - y + z &= 3 \\ 2x + 2y - 3z &= 7 \\ x - 2y + z &= 8 \\ x + 2y + 11z &= 49 \end{aligned}$$

5. Solve:
$$\begin{aligned} x - y + z &= 3 \\ 2x + 2y - 3z &= 7 \\ x - 5y + 6z &= 2 \\ 11x - 7y + 6z &= 34 \end{aligned}$$

6. Solve:
$$\begin{aligned} x - y + z &= 3 \\ 2x + 2y - 3z &= 7 \\ x - 2y + z &= 8 \\ 2x + y - z &= 2 \end{aligned}$$

7. Solve:
$$\begin{aligned} x + 3y &= 6 \\ 3x - y &= 5 \\ -x + 7y &= 7 \\ 2x - 4y &= -1 \end{aligned}$$

8. Solve:
$$\begin{aligned} x + 2y + 3z - 6w &= 12 \\ -x + y - 2z + 6w &= 9 \\ x + 5y + 4z - 6w &= 30 \end{aligned}$$

9. Solve:
$$\begin{aligned} x - y + z + w &= 4 \\ x + 2y - 3z + w &= 3 \\ x - y + z &= 6 \end{aligned}$$

10. Solve:
$$\begin{aligned} x + 2y + 3z - 6w &= 12 \\ -x + y - 2z + 6w &= 9 \\ x + 5y + 4z - 6w &= 33 \end{aligned}$$

3.3 Solutions of Systems Using Tableaus

To save writing and to make solution of systems possible on a computer (which cannot do algebra), an ingenious technique has been developed. This method involves suppressing the symbols for the unknowns and recording only their coefficients while performing the allowable operations on the equations. In short, a tableau or "matrix" of the coefficients in the system is constructed. For example, corresponding to the system

$$\begin{aligned} x - 3y + z &= 6 \\ 2x + 3y &= 9 \\ x + 15y - z &= 4 \end{aligned}$$

the tableau is:

$$\left(\begin{array}{ccc:c} 1 & -3 & 1 & 6 \\ 2 & 3 & 0 & 9 \\ 1 & 15 & -1 & 4 \end{array}\right) \quad \begin{array}{l} \text{Row 1} \\ \text{Row 2} \\ \text{Row 3} \end{array}$$

The "rows" are the coefficients of x, y and z and the righthand member of each equation in turn. Note that the numbers are carefully lined up in columns so that one can always tell to which unknown the coefficient is attached. For this reason, it is necessary to enter a 0 whenever an unknown is not present. Rows are numbered from top to bottom and the columns are numbered from left to right. Thus, one may refer to the entry in Row 2, Column 3, which in this case is 0. Thus in the second equation, the coefficient of the third unknown is zero.

To solve the system, one may perform all the allowable operations on equations on the rows of the matrix. That is,

1. any row may be multiplied by any nonzero number. Multiplying a row by a number thus means multiplying every entry by that number.
2. any two rows may be interchanged.
3. any multiple of one row may be added to any other row. Adding two rows means, of course, adding the elements in like columns.

A new matrix obtained by any of these operations or any combination of them is said to be "row equivalent" to the original. That is, row equivalent matrices represent equivalent systems. Let us attack the system given at the beginning of this section by this method. In solving equations, one first uses the first equation to eliminate the first unknown from all succeeding ones. In matrix form, this means use the first row to make all entries below it in the first column zero. That is, replace Row 2 by Row 2 minus twice Row 1 and replace Row 3 by Row 3 minus Row 1:

$$\left(\begin{array}{ccc|c} 1 & -3 & 1 & 6 \\ 0 & 9 & -2 & -3 \\ 0 & 18 & -2 & -2 \end{array}\right)$$

Next, we clear the second unknown from all equations beyond the second, i.e., we clear everything below Row 2 in the second column. To do this, we double Row 2 and subtract it from Row 3, which yields

$$\left(\begin{array}{ccc|c} 1 & -3 & 1 & 6 \\ 0 & 9 & -2 & -3 \\ 0 & 0 & 2 & 4 \end{array}\right)$$

We have now completed the triangularization. The third row reads

$$2z = 4$$

so $z = 2$. The second row reads

$$9y - 2z = -3$$

since $z = 2$, $9y - 4 = -3$, $9y = 1$, $y = 1/9$. The first row reads

$$x - 3y + z = 6$$

Substituting, we have

$$x - 1/3 + 2 = 6 \qquad x = 13/3$$

We shall use the symbol $\sim$ to indicate equivalence. Let us attempt to solve the following:

$$\begin{aligned} x + y + z &= 2 \\ x + z &= 4 \\ y - z &= -5 \\ x - 2y - z &= 5 \end{aligned}$$

which in matrix form is

$$\left(\begin{array}{rrr:r} 1 & 1 & 1 & 2 \\ 1 & 0 & 1 & 4 \\ 0 & 1 & -1 & -5 \\ 1 & -2 & -1 & 5 \end{array}\right) \sim \left(\begin{array}{rrr:r} 1 & 1 & 1 & 2 \\ 0 & -1 & 0 & 2 \\ 0 & 1 & -1 & -5 \\ 0 & -3 & -2 & 3 \end{array}\right)$$

$$\sim \left(\begin{array}{rrr:r} 1 & 1 & 1 & 2 \\ 0 & -1 & 0 & 2 \\ 0 & 0 & -1 & -3 \\ 0 & 0 & -2 & -3 \end{array}\right) \sim \left(\begin{array}{rrr:r} 1 & 1 & 1 & 2 \\ 0 & -1 & 0 & 2 \\ 0 & 0 & -1 & -3 \\ 0 & 0 & 0 & 3 \end{array}\right)$$

The triangularization is complete and the last row reads $0 = 3$. Again, we have an incompatible system. Were you able to fill in the operations as we went from matrix to matrix? Let us try it again only change the last equation to

$$x - 2y - z = 2$$

This will change the matrix only in that the entry in Row 4, Column 4, will be 2 instead of 5. Now

$$\left(\begin{array}{ccc|c} 1 & 1 & 1 & 2 \\ 1 & 0 & 1 & 4 \\ 0 & 1 & -1 & -5 \\ 1 & -2 & -1 & 2 \end{array}\right)$$

and we begin by clearing the first column,

$$\left(\begin{array}{ccc|c} 1 & 1 & 1 & 2 \\ 0 & -1 & 0 & 2 \\ 0 & 1 & -1 & -5 \\ 0 & -3 & -2 & 0 \end{array}\right) \quad \begin{array}{l} \\ \text{Row 2} - \text{Row 1} \\ \text{Unchanged} \\ \text{Row 4} - \text{Row 1} \end{array}$$

Next we clear Column 2,

$$\left(\begin{array}{ccc|c} 1 & 1 & 1 & 2 \\ 0 & -1 & 0 & 2 \\ 0 & 0 & -1 & -3 \\ 0 & 0 & -2 & -6 \end{array}\right) \quad \begin{array}{l} \\ \\ \text{Row 3} + \text{Row 2} \\ \text{Row 4} - 3 \text{ times Row 2} \end{array}$$

Finally we clear Column 3,

$$\left(\begin{array}{ccc|c} 1 & 1 & 1 & 2 \\ 0 & -1 & 0 & 2 \\ 0 & 0 & -1 & -3 \\ 0 & 0 & 0 & 0 \end{array}\right) \quad \begin{array}{l} \\ \\ \\ \text{Row 4} - 2 \text{ times Row 3} \end{array}$$

The equations now read

$$\begin{aligned} x + y + z &= 2 \\ -y \quad\quad &= 2 \\ -z &= -3 \\ 0 &= 0 \end{aligned}$$

The equations are compatible and solving by back substitution yields

$$z = 3 \qquad y = -2 \qquad x = 1$$

EXERCISES

1–8. Redo the problems in Sec. 3.1 in matrix form.

9–18. Redo the problems in Sec. 3.2 in matrix form.

19. Solve:

$$\begin{aligned} x - y + z &= 6 \\ x + 2y + z &= 11 \\ 3x \phantom{{}+13y} + 3z &= 23 \\ 2x + 13y + 2z &= 37 \\ -x + 13y - z &= 14 \end{aligned}$$

4

Introduction to Probability and Game Theory

4.1 Basic Ideas

The origin of the modern theory of probability is reputedly in a question raised by a gambler, the Chevalier de Mere, to the French mathematician and philosopher, Blaise Pascal (1623–1662). The question supposedly concerned how to divide the stakes in a game which is interrupted before completion. As an illustration, consider the following game. Two players, A and B, agree to play a dice game. Each player will roll one die. The one who rolls the higher number will collect as many dollars as the sum of the two dice. Player A rolls his die and gets a two. At this point the game is stopped. Clearly, the situation as it stands seems to favor Player B. If the game can never be resumed, how much should A pay to B? Another way of asking the same question is, How much would a bystander be willing to pay to B for the right of taking his place in the game at this juncture? It seemed to the mathematicians that the fair way to settle the game would be for A to pay to B an amount equal to what he could honestly expect to win on the average whenever this situation occurred. Average value could be obtained as follows: First, assume that the dice are perfect balanced cubes. Then if B were to roll his die a large number of times, one would expect that he would get each of the numbers one, two, three, four, five and six, each about one-sixth of the time.

Table 1 summarizes the results of these outcomes.

Table 1

B's Number	B's Winnings	Fraction of Time
1	-3	1/6
2	0	1/6
3	5	1/6
4	6	1/6
5	7	1/6
6	8	1/6

So we would expect that B will win -3 (i.e., lose 3), 0, 5, 6, 7 or 8 dollars, each about one-sixth of the time. Hence, on the average, B may expect to win $(-3 + 0 + 5 + 6 + 7 + 8)/6 = 23/6$ dollars every time this situation occurs. Note that under no condition can the sum of 23/6 of a dollar actually be the amount won. Nevertheless, this is the *average* value of the game to B and this is the amount for which he should be willing to sell his turn. In the development of the theory of probability, an attempt is made to set up a formal mathematical structure for problems of this type and others of a related nature.

Let us begin by assuming that an experiment is about to be performed. The experiment may be rolling a die, or flipping a coin, or measuring the height of a person selected from names in the telephone book by jabbing a pin into a page or it might be simply observing whether rain will be falling at noon tomorrow in Chicago. However, one characteristic is necessary, that is, the outcome of the experiment must be at least partly determined by chance so that there is no way to predict with certainty the exact outcome. Let us further assume that there are only a finite number of possible outcomes to the experiment. When a coin is flipped, the possible outcomes are "heads" and "tails," assuming that the coin will not land on its edge. Of course, one could observe the angle between some axis on the coin and true north, or one could observe the temperature of the room after the coin is flipped, but these seem to have little bearing in the usual case.

Let us assume that one could write a list of the outcomes of interest. Also, assume that this list is finite, that all the possible outcomes are included and that anything that can occur is described by precisely one item on the list. Such a list is called a set of *elementary outcomes* and the Universe S, composed of these outcomes, is called a *sample space* for the experiment. For example, when a die is rolled, $S = \{1, 2, 3, 4, 5, 6\}$ gives a sample space for the experiment. In measuring the height of a person selected at random, we would not get a sample space as defined above, since one cannot list all the possible heights. If, on the other hand, one agrees to measure to the nearest inch, then we can probably assume that $S = \{1, 2, \ldots, 120\}$ is a reasonable sample space, since people over 10 feet tall are rather scarce.

Of course, one may be interested in something other than an elementary outcome. For example, one may wish to describe the case in which B wins in the game described at the outset. Clearly, B wins if he rolls a three, four, five or six. Letting $W = \{3, 4, 5, 6\}$, we see that the outcome "B wins" can be described by the words "one of the elements in W occurs" or simply "W occurs." W is an example of a composite outcome or "event," that is, an event is a subset of the sample space. An event is said to occur if any element in the event is the outcome of the experiment. Let us look at some simple examples. In rolling two dice, one is usually interested in the sum of the two dice. Therefore, we shall take as the sample space the set $S = \{2, 3, 4, \ldots, 12\}$. The event, "a number divisible by 3 occurs," $= A = \{3, 6, 9, 12\}$. The event, "a number bigger than 5 and smaller than 10 occurs," $= B = \{6, 7, 8, 9\}$.

Common English sentences describing the occurrence or nonoccurrence of certain events now take on their natural set notational interpretation. Thus, if a person says, "I'll bet that a number divisible by 3, falling between 5 and 10, will be rolled," he is betting that A and B will occur simultaneously. That is, in set notation, he is betting on $A \cap B = \{6, 9\}$. Similarly, the statement "a number between 5 and 10 or divisible by 3" is interpreted as $A \cup B = \{3, 6, 7, 8, 9, 12\}$. To say that A does not occur is to refer to A'. Since either A or A' must occur, we say that $A \cup A'$ is a "sure" or "certain" event. That is, $A \cup A' = S$, the sample space, is sometimes referred to as the sure event. Similarly, since A and A' cannot both occur at once, $A \cap A'$ is called the *impossible event*. So, the symbol $\varnothing$, used for the empty set, will signify, in probability theory, an event that cannot occur.

Interestingly, it is possible to construct many different sample spaces for the same experiment. In fact, given any sample space for an experiment, any partition of the sample space into disjoint sets which together encompass all outcomes will serve as a new sample space. For example, suppose an experiment consists of selecting a card from a well-shuffled deck of playing cards. The usual sample space would consist of 52 points identified by suit and rank. Thus, Ca = club ace; Hq = heart queen; S7 = spade seven and Dk = diamond king. However, one could just as easily deal with the sample space $S_1 = \{\text{Heart, Spade, Diamond, Club}\}$. Certainly any card selected falls into one of these categories. No card falls into two categories simultaneously and so all the conditions for a set of elementary outcomes are satisfied. On the other hand, $S = \{\text{Heart, Face card, Diamond, Black card}\}$, is not a proper sample space. Any card picked will surely fall into one of the categories, but the Queen of Hearts, for example, falls into two at once. In some problems it is difficult to know whether all possibilities are included or whether they are mutually exclusive. However, in most of what we do this should not offer any difficulty.

EXERCISES

1. Calculate the value to B of the dice game described at the beginning of this section, if A had rolled a 1. Now try it for the other possible numbers that A could have rolled. Find the average of the values for the cases when A rolled 1, 2, 3, 4, 5 and 6. Does this result surprise you?

2. In playing certain board games, two dice are rolled, the sum of the two dice tells how far to move and a roll of doubles gives an extra turn. Write a sample space suitable for the experiment of rolling the dice in these games.

3. A deck consists of ten cards numbered 1 to 10. A card is drawn at random from the deck. Write a sample space for this experiment.

4. Write a sample space consisting of five elements for the experiment in Exercise 3 and such that (a) you think the elementary outcomes are equally likely to occur, and (b) you think the outcomes are not equally likely.

5. A bag contains balls lettered $a, b, \ldots, z$. A ball is chosen at random from the bag. Let the sample space be $S = \{a, b, c, \ldots, z\}$. Write the event, V, a vowel is drawn. Write the event, E, a letter in the word *event* is drawn. How would you express the event $\{e\}$ in terms of E and V? What is the event E or V occurred? What is the event E occurred but V did not?

6. A bag contains four blocks colored red, green, blue and yellow. A block is selected at random. Let the sample space be the four possible colors. Write all possible events for this experiment.
7. A red die and a green die are rolled. It is important to know what numbers turned up on each die. Write a suitable sample space for this experiment.
8. A man has five nails which he is trying to drive into a board. The nails bend easily. He tries each one in turn until he succeeds. What is the sample space showing all possible outcomes for this experiment?
9. For the experiment described in Exercise 8, display the following events: It takes at least three tries to succeed; it takes no more than three tries to succeed; either it works the first time or not at all; it does not work at first but at least one nail is left over.
10. Weather observers in New York and Chicago report either clear weather, rain or snow. Write the set of all possible joint communiques. Write the following events: It is raining in at least one city; it is raining in exactly one of the two cities; the weather is the same in both cities; the weather is not the same in both cities; it is snowing in New York; it is snowing in one city and not in the other.

4.2 Probability Measure

In all of the experiments described and in any other one that may be encountered, it is clear that some events seem to be more likely to occur than others. For example, if $A \subset B$ then whenever A occurs so does B, but B may occur without A doing so. Hence, B is more likely than A. Or, as a matter of experience, gamblers have noted that in rolling two dice, one is far more likely to get a total of seven than of two. In the theory of probability, mathematicians attempt to apply a quantitative measure of likelihood to the possible subsets (events) of a sample space. It would be nice to express these likelihoods in terms of percents. That is, a sure event has 100% chance of occurring and an impossible event has 0% chance of occurrence. However, in order to avoid percent signs, it has been agreed to use numbers on the scale of zero to one. That is, a likelihood of one means certainty and zero means impossibility. (In general probability theory this is not quite true, but it is correct for finite probability theory, which is the only type to be discussed in this text.)

In order to have this measure of likelihood in accordance with intuition, we shall define the "probability" of an event E, $P(E)$, to have the following properties:

1. If E is any event, $0 \leq P(E) \leq 1$.
2. $P(S) = 1$.
3. If $A \cap B = \varnothing$, then $P(A \cup B) = P(A) + P(B)$.

These "Probability Axioms" have a very simple intuitive basis. Axiom 1 simply states that the measure of likelihood shall be scaled from zero to one. Axiom 2 states that the likelihood of a sure event is one and hence fixes the top end of the scale. Axiom 3 says that if two events cannot occur simultaneously, then the probability of one or the other occurring is simply the sum of the probabilities of the events separately occurring. Perhaps a simple example would illustrate this. Let a card be selected from a deck at random. Let H be the event a heart is drawn and D be the event a diamond is drawn. One would suspect that $P(H) = 1/4$ and $P(D) = 1/4$ since there are four suits and no one suit seems more likely to be drawn than the other. Now, $P(\text{Red card}) = P(H \cup D)$. Since H and D are mutually exclusive, by Axiom 3, $P(H \cup D) = P(H) + P(D) = 1/4 + 1/4 = 1/2$. The final result also seems to agree with intuition.

In applying Axiom 3, one must be certain that the events are mutually exclusive. For example, suppose a coin is tossed twice. Let A = a head occurs on the first toss and B = a head occurs on the second toss. Consider $P(\text{Head on either first or second toss}) = P(A \cup B)$. If the coin is fair, it seems reasonable to attach probability 1/2 to both A and B. If one were now to apply Axiom 3, the result would be $P(A \cup B) = P(A) + P(B) = 1/2 + 1/2 = 1$. This means that in flipping a coin twice, one is certain to get at least one head. Of course, experience tells us that this is not at all the case. The error enters in, in that A and B are not mutually exclusive, that is, the probability of getting two heads in a row is not zero. In fact, we shall see a little later that in two independent flips of a fair coin, the probability of getting two heads is 1/4. We shall also discover how to use this information to obtain a proper result for the preceding example. However, let us first investigate how one goes about constructing probability measures on the events of a finite sample space.

Let $S = \{s_1, s_2, s_3, \ldots, s_n\}$ be a finite sample space for some experiment. A probability measure may be applied to all the subsets of S as follows: Associate with each point s_j a number p_j which measures the likelihood of s_j. The only restrictions on the p's are

1. $p_j \geq 0$ for all $j = 1, 2, \ldots, n$;
2. $p_1 + p_2 + \cdots + p_n = 1$.

We shall define the probability of any set A, $P(A)$, as the sum of the likelihoods of the points in A. In particular, the probability of the singleton set $\{s_j\}$ is just p_j. Hence, we shall mix our notation and sometimes refer to the likelihood of a point as the probability of the point and thus not really distinguish between the point s and the set $\{s\}$ whose only element is s. To a set theorist this is an unforgivable sin, but for our purposes there will be little chance of confusion.

Suppose, for example, that we have a rather biased coin with a wide edge. In flipping this coin we have three possible outcomes, h, t and e. Let us assign the likelihood $1/2$ to h, $1/3$ to t and $1/6$ to e. The three likelihoods are nonnegative numbers whose sum is 1. This experiment has eight possible events associated with it, namely,

1. $H =$ Heads occurs; $P(H) = 1/2$;
2. $T =$ Tails occurs; $P(T) = 1/3$;
3. $E =$ Coin lands on edge; $P(E) = 1/6$;
4. $H' = \{t, e\} =$ Heads does not occur; $P(H') = 1/3 + 1/6 = 1/2$;
5. $T' = \{h, e\} =$ Tails does not occur; $P(T') = 1/2 + 1/6 = 2/3$;
6. $E' = \{h, t\} =$ The coin does not land on its edge; $P(E') = 1/2 + 1/3 = 5/6$;
7. $S = \{h, t, e\} =$ The coin lands in one of the three possible positions; $P(S) = 1$;
8. $\varnothing =$ The coin lands in none of the three positions; $P(\varnothing) = 0$.

The observant reader will have noticed in this example that the probability of $\varnothing$, the impossible event, is 0. We have remarked earlier that this should be the case, but have not stated it as one of our axioms. We shall prove this as our first result on probability.

THEOREM 1 Let S be the sample space for any experiment; then $P(\varnothing) = 0$.

Proof From Axiom 2, $P(S) = 1$. Now $S \cup \varnothing = S$ and $S \cap \varnothing = \varnothing$, so by Axiom 3, $P(S) = P(S) + P(\varnothing)$. Therefore, $1 = 1 + P(\varnothing)$; $P(\varnothing) = 0$. The method of this proof may be extended to the following more general result.

THEOREM 2 Let S be the sample space for any experiment and let A be any subset of S; then, $P(A') = 1 - P(A)$.

Proof $A \cup A' = S$ and $A \cap A' = \varnothing$. Hence $1 = P(S) = P(A \cup A')$. Therefore, $1 = P(A) + P(A')$ and $P(A') = 1 - P(A)$.

The reader will note that this theorem does indeed hold for the example discussed above; but he should also realize that the proof does not depend upon the method by which the probability measure is attached to the subsets of S but only upon the three axioms. Let us consider one other well-known result, which deals with the probability of the union of sets which are not disjoint.

THEOREM 3 Let S be the sample space of an experiment and let A and B be any two subsets of S; then $P(A \cup B) = P(A) + P(B) - P(A \cap B)$.

Proof Referring to chapter 1, figure 1c, we can write $A \cup B = A \cup (A' \cap B)$ and $A \cap (A' \cap B) = \varnothing$. Therefore, $P(A \cup B) = P(A) + P(A' \cap B)$. Furthermore, $B = (A \cap B) \cup (A' \cap B)$ and $(A \cap B) \cap (A' \cap B) = \varnothing$. Therefore, $P(B) = P(A \cap B) + P(A' \cap B)$. This gives two basic relationships:

$$P(A \cup B) = P(A) + P(A' \cap B) \tag{1}$$

$$P(B) = P(A \cap B) + P(A' \cap B) \tag{2}$$

Subtracting Eq. (2) from Eq. (1), we have

$$P(A \cup B) - P(B) = P(A) - P(A \cap B)$$

or

$$P(A \cup B) = P(A) + P(B) - P(A \cap B)$$

which completes the proof.

How does one go about understanding the intuitive justification of such a result? There is a very nice geometric method for visualizing probabilistic results. Let the sample space be represented by a square of area 1. Let all the events be subregions of this square just as in a Venn diagram. Let the probability of an event be the area of the subregion. Referring to figure 1, we are looking for the area of $A \cup B$. The area of A plus the area of B is clearly larger than the area of $A \cup B$. In fact, it seems obvious that the excess is caused by the fact that we have added in the area of $A \cap B$ twice. Hence, area of $A \cup B$ = area of A + area of B − area of $A \cap B$.

When this type of geometric reasoning is used, it is apparent that if $A \subset B$, then the area of A is less than or equal to the area of B. Thus, we should be able to prove the following theorem.

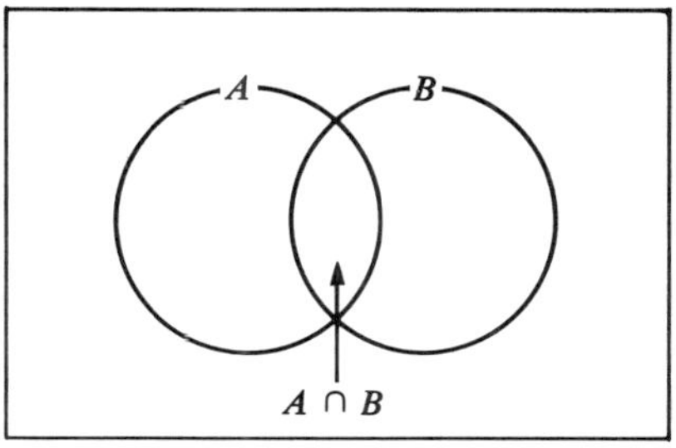

Figure 1

THEOREM 4 Let A and B be events and let $A \subset B$. Then $P(A) \leq P(B)$.

Proof $B = A \cup (A' \cap B)$, where $A \cap (A' \cap B) = \varnothing$. Thus, $P(B) = P(A) + P(A' \cap B)$. Since $P(A' \cap B) \geq 0$, by Axiom 1 it follows that $P(B) \geq P(A)$.

Let us demonstrate the use of some of these theorems with an example. A certain political party is backing three candidates for different offices. Their names are Adams, Bailey and Calhoun. The pollsters tell us that Adams has a 34% chance of winning, Bailey has a 44% chance of winning and there is a 21% chance that both will be elected. What is the probability that at least one will be a winner? Let A be the event Adams wins, and let B be the event Bailey wins. Then

$$P(A) = 0.34 \qquad P(B) = 0.44 \qquad P(A \cap B) = 0.21$$

Therefore,

$$\begin{aligned} P(A \cup B) &= P \text{ (at least one wins)} \\ &= P(A) + P(B) - P(A \cap B) \\ &= 0.34 + 0.44 - 0.21 = 0.57 \end{aligned}$$

Let us assume additionally that the probability that both Bailey and Calhoun are elected is 0.20 and the probability that at least one is elected is 0.79. What is the probability that Calhoun is elected?

Let C be the event that Calhoun is elected. Then

$$\begin{aligned} P(B \cup C) &= P(B) + P(C) - P(B \cap C) \\ 0.79 &= 0.44 + P(C) - 0.20 \\ P(C) &= 0.55 \end{aligned}$$

Suppose that the probability that all three are simultaneously elected is 0.12. What is the probability that Bailey and Calhoun get elected but Adams does not? Here we are given $P(A \cap B \cap C) = 0.12$ and asked to find $P(A' \cap B \cap C)$. Now $B \cap C = (A \cap B \cap C) \cup (A' \cap B \cap C)$, and $A \cap B \cap C$ and $A' \cap B \cap C$ are disjoint so we can apply Axiom 3 directly:

$$P(B \cap C) = P(A \cap B \cap C) + P(A' \cap B \cap C)$$
$$0.20 = 0.12 + P(A' \cap B \cap C)$$
$$P(A' \cap B \cap C) = 0.08$$

If $P(A \cap B' \cap C) = 0.07$, what is $P(A \cup C)$?

$$P(A \cup C) = P(A) + P(C) - P(A \cap C)$$

and

$$P(A \cap C) = P(A \cap B' \cap C) + P(A \cap B \cap C)$$
$$= 0.07 + 0.12 = .19$$
$$P(A \cup C) = 0.34 + 0.55 - 0.19 = 0.70$$

What is the probability that none are elected? This could be done directly, but figure 2 shows how all the information given previously can be summarized to completely expose the answers to any problems about this example.

The last part of the preceding example really requires finding $P(A \cup B \cup C)$, where A, B and C are not mutually exclusive. There is a general formula for this. In fact, there is a general formula for $P(A_1 \cup A_2 \cup A_3 \ldots \cup A_n)$. For now we content ourselves with the following theorem.

THEOREM 5 $P(A \cup B \cup C) = P(A) + P(B) + P(C) - P(A \cap B) - P(A \cap C) - P(B \cap C) + P(A \cap B \cap C)$.

The proof of this result is the first exercise. However, one may convince oneself of this result geometrically by means of a Venn diagram.

EXERCISES

1. Prove Theorem 5.
2. Referring to Exercise 3, Section 4.1, assume all cards are equally likely. Let X be the number on the card drawn. Let $A = X$ is even and let $B = 2 \leq X < 8$. Find $P(A)$, $P(B)$, $P(A \cap B)$ and $P(A \cup B)$.

3. Referring to Exercise 5, Section 4.1, assume all the letters are equally likely to be drawn. Find $P(V)$, $P(E)$, $P(E \cup V)$ and $P(E \cap V')$.

4. Referring to Exercise 6, Section 4.1, suppose that red is twice as likely as green which is twice as likely as blue which is just as likely as yellow. Find the probabilities of all the events for this sample space.

5. Referring to Exercise 8, Section 4.1, let X be the number of the attempt on which the first success occurs. Suppose $P(X = 1) = 1/3$, $P(X = 2) = 2/9$, $P(X = 3) = 2/9$, $P(X = 4) = 1/9$ and $P(X = 5) = 1/18$. Find the probabilities of all the events in Exercise 9.

6. Of students studying French and English, 95% pass at least one of the courses, 78% pass French and 88% pass English. What percent pass both?

7. George and Harry are unreliable workers; the probability that at least one of them will be out sick is 0.6. The probability that Harry will be out sick is 0.5 and the probability that they will both be out is 0.2. What is the probability that George will be out? What is the probability that neither is sick? What is the probability that exactly one will be sick? What is the probability that George will be sick and Harry will not be?

8. Continuation of Exercise 7. Sam is also unreliable and counting him, one of the three men is out 90% of the time. The probability that only Sam is sick is 0.2. Find the probability that Sam and at least one of the others are out sick.

9. A survey of luncheonettes shows that in addition to conducting their regular business many of them sell reading matter as well. In fact, 1/2 sell newspapers, 23/36 sell magazines, 2/3 sell paperback books, 7/18 sell newspapers and magazines, 1/3 sell magazines and paperbacks, 10/36 sell newspapers and paperbacks, and 1/6 sell all three. (a) What fraction sell no reading matter? (b) What fraction sell only newspapers? (c) What fraction sell exactly one of the three types of reading matter?

10. A recent advertising survey has shown that a person selected at random watches TV commercials with probability 0.8, reads newspaper ads with probability 0.7 and reads magazine ads with probability 0.4. Only 10% of the people do none of these things and 20% do all three. Find the probability that a person selected at random is exposed to exactly two of these three forms of advertising.

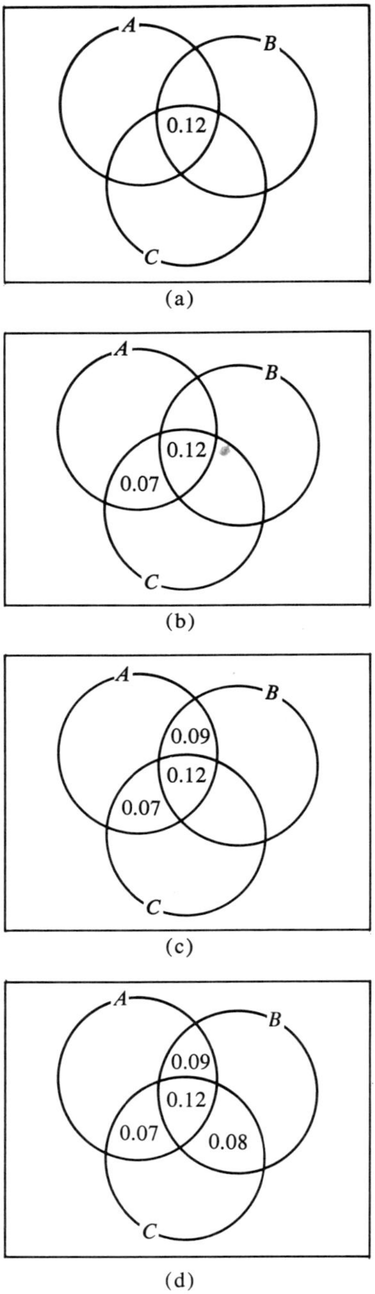

(a) $P(A \cap B \cap C) = 0.12$

(b) $P(A \cap C) = 0.19$

(c) $P(A \cap B) = 0.21$

(d) $P(B \cap C) = 0.20$

Figure 2

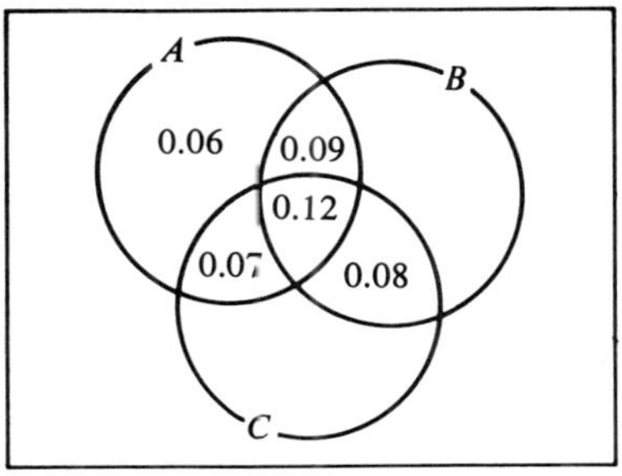

(e)

$P(A) = 0.34$

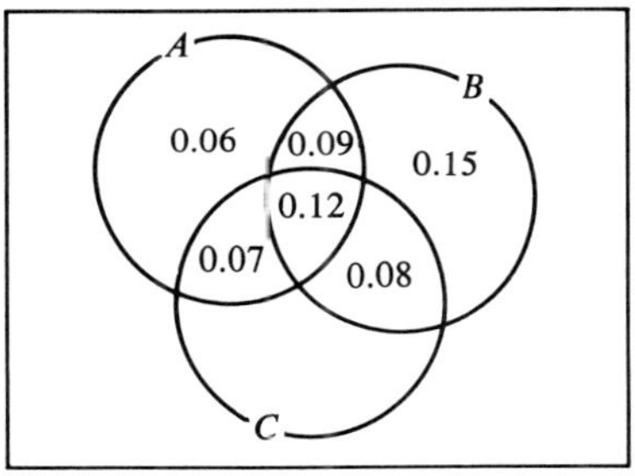

(f)

$P(B) = 0.44$

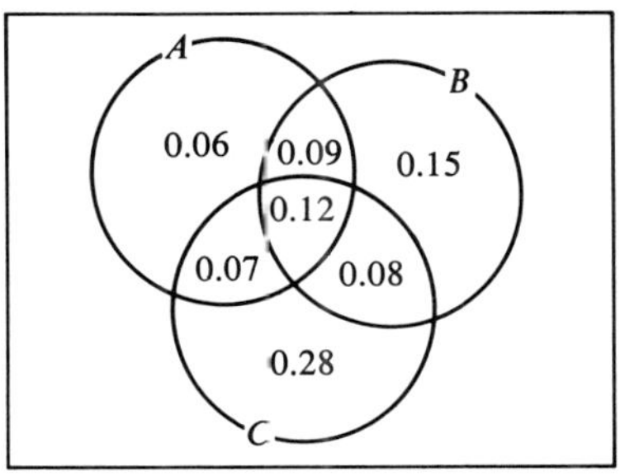

(g)

$P(C) = 0.55$

$P(A \cup B \cup C) = 0.06 + 0.09 + 0.07 + 0.12 + 0.15 + 0.08 + 0.28 = 0.85$

$P(A \cup B \cup C) = 0.85$

$P([A \cup B \cup C]') = 1 - P(A \cup B \cup C) = 0.15$

Figure 2, Cont.

4.3 Assigning Probabilities

Now that we know how to handle combinations of events given the probabilities of the component parts, there still remains some doubt as to how one begins to assign probability measures to a completely new sample space. Certainly, the most straightforward approach seems to be as follows: Describe and list all the possible elementary outcomes for the experiment and assign to each one a likelihood as described in the preceding section. The difficulty arises in assigning likelihoods in such a way as to mirror reality, that is, to pick the likelihood so that the likelihood of a point s is the fraction of the time that one would really expect s to occur. To be more precise, let us suppose that we are able to repeat an experiment a large number of times under essentially identical conditions and in such a way that the occurrence or nonoccurrence of an event in any trial is totally independent of what happened previously. For example, in flipping a coin, one feels relatively confident that each flip is unaffected by the others, that the result of each flip is random, and that there are no outside influences that could affect the likelihood of a head or a tail.

Now, each time the experiment is performed, there is a sample space $S = \{s_1, s_2, s_3, \ldots, s_k\}$. If the experiment is performed N times, then s_1 will have occurred n_1 times, s_2 will have occurred n_2 times and so on. We would then be able to record the frequency with which each elementary outcome occurred. That is, $f_1 = n_1/N, f_2 = n_2/N, \ldots, f_k = n_k/N$. Note that $n_1 + n_2 + \cdots + n_k = N$, or, in other words, $f_1 + f_2 + \cdots + f_k = 1$. This much can always be done. Whether or not these frequencies mean anything is a philosophical question. However, we do assume that there is some regularity in even a probabilistic universe. We exhibit our faith in this proposition via the so-called Principle of Statistical Regularity. This states that there exist numbers $p_1, p_2, \ldots, p_k$ such that for large values of N, f_j is approximately p_j for $j = 1, 2, 3, \ldots, k$. These p's are the likelihoods of the outcomes $s_1, s_2, \ldots, s_k$. Therefore, when faced with a new experiment, one should perform a large number of trials and measure the frequencies of the possible outcomes. These frequencies may then serve as a first approximation to the likelihoods of the elementary outcomes. One may then define the probability of each event in the sample space.

As a simple example, let us consider rolling a single die. An actual string of outcomes in successive trials yielded 5, 4, 4, 1, 2, 1, 1, 2, 3, 6, 3, 2, 1, 3, 2, 1, 2, 5, 4, 6. We may now display the result of these twenty trials. (See table 2.)

Of course, this is a very small number of trials. A second string of twenty yielded 5, 5, 6, 1, 1, 3, 4, 6, 2, 3, 2, 3, 5, 3, 5, 3, 3, 4, 2, 6. The results of combining these with the first group are given in table 3.

Table 2

20 Trials

No. of 1's = 5	$f_1 = 5/20 = .25$
No. of 2's = 5	$f_2 = 5/20 = .25$
No. of 3's = 3	$f_3 = 3/20 = .15$
No. of 4's = 3	$f_4 = 3/20 = .15$
No. of 5's = 2	$f_5 = 2/20 = .10$
No. of 6's = 2	$f_6 = 2/20 = .10$

Table 3

40 Trials

No. of 1's = 7	$f_1 = 7/40 = .175$
No. of 2's = 8	$f_2 = 8/40 = .200$
No. of 3's = 9	$f_3 = 9/40 = .225$
No. of 4's = 5	$f_4 = 5/40 = .125$
No. of 5's = 6	$f_5 = 6/40 = .150$
No. of 6's = 5	$f_6 = 5/40 = .125$

Continuing, we performed 160 more trials, reaching a total of 200 with the results in table 4.

Table 4

200 Trials

No. of 1's = 27	$f_1 = .135$
No. of 2's = 37	$f_2 = .185$
No. of 3's = 39	$f_3 = .195$
No. of 4's = 28	$f_4 = .140$
No. of 5's = 35	$f_5 = .175$
No. of 6's = 34	$f_6 = .170$

Using these figures, one might say, for example,

$$P[\text{Even number}] = P(E) = P\{2, 4, 6\} = f_2 + f_4 + f_6 = 0.495$$

Of course, there is another point of view. Namely, since a die is totally symmetric, no face should be favored over any other, if it is well made. Therefore, the likelihood of each elementary outcome should be taken to be 1/6. From this angle, the data collected so far may be viewed as an attempt to ascertain whether the actual die rolled is "fair," that is, whether the real die behaved as a mathematically ideal die should. Answers to questions of this type fall into the province of statistical theory.

Nevertheless, the idea of a mathematically ideal die, or fair coin or "perfectly shuffled" deck of cards is pervasive in elementary probability theory. It is expressed as the "Principle of Equal Likelihood." This principle says that in the absence of evidence to the contrary, one should assume that all elementary outcomes have equal likelihood. That is, if an experiment has k possible elementary outcomes, each should be assigned the probability $1/k$ unless experience dictates otherwise. Now, look back at the table of 200 trials (table 4). Do you think that the frequencies $f_1, f_2, \ldots, f_6$ differ significantly from $1/6 = 0.167$? We will not attempt an answer here.

Let us consider another example. In rolling two dice, we will record only the sum of the faces which come up. If we consider our sample space to be $S = \{2, 3, 4, \ldots, 12\}$ and assign the probability $1/11$ to each point, we would have a perfectly valid probability measure established in accordance with the principle of equal likelihood. However, gamblers have ascertained over the centuries that this is a very poor model of reality. In fact, when two dice are rolled, a total of seven will occur about six times as often as a total of two. Interestingly enough, this disparity can be cleared up within the framework of the equal likelihood principle by choosing the elementary outcomes more critically. To be exact, one should consider two dice of different color (so that they may be easily distinguished) and consider the elementary outcomes to be all possible distinguishable outcomes. That is, calling the dice red and green, there are 36 outcomes, ranging from $(R1, G1)$ to $(R6, G6)$, each of which may be assigned likelihood $1/36$. We leave the actual listing of all the outcomes to the reader. As an example, however, $P(\text{Total of the dice is } 7) = P\{(R1, G6), (R2, G5), (R3, G4), (R4, G3), (R5, G2), (R6, G1)\} = 6/36 = 1/6$.

Thus, the equal likelihood method of assigning probabilities is momentarily revived. Still, in flipping biased coins or rolling "loaded" dice, the principle is of no use. Nevertheless, we shall use it repeatedly. Whenever we refer to flipping a fair coin, we mean $P(H) = P(T) = 1/2$. Rolling a fair die means $P(1) = P(2) = \cdots = P(6) = 1/6$. Or, in general, the phrase, "select an object at random from a set," will mean that all objects in the set are to be equally likely selections.

EXERCISES

1. List all the outcomes in rolling a red die and a green die. Find the probabilities of each of the totals $2, 3, \ldots, 12$ assuming each elementary outcome has probability $1/36$.

2. List all the outcomes possible when a coin is flipped three times in a row. Assuming heads and tails are equally likely, find
   (a) *P* (Three heads)
   (b) *P* (At least two tails)
   (c) *P* (Odd number of heads)
   (d) *P* (Last flip is heads)
   (e) *P* (First or last flip is tails)
   (f) *P* (First and second flip are tails)
   (g) *P* (At least one head)

3. Flip a penny and a dime simultaneously. Repeat 100 times. Find the frequency of two heads, one head, two tails, head on the penny and tail on the dime.

4. Radioactive decay simulation. Let ten pennies represent ten atomic particles. Flip the ten coins and discard all those that come up heads. Flip the remaining coins again; discard all that come up heads. Continue in this manner until no coins remain. Each flip represents a half-life. The number of flips needed is the number of half-lives until total decay. Repeat the experiment 100 times and find the frequencies of one flip, two flips, etc.

5. A poker player has four hearts and a spade. He discards the spade and draws one card. What is the probability that he ends up with a heart flush? HINT: Since he does not know his opponents' cards, assume all 47 remaining cards are equally likely.

6. A poker player has two pairs. He discards his odd card and draws one. What is the probability that he gets a full house?

7. Continuation of 5 and 6. What is the probability of filling an "inside straight," a straight open at both ends? Drawing to *H*5, *H*6, *H*7, *H*8, what is the probability of ending up with a straight, a flush, or a straight flush?

4.4 Random Variables

In many mathematical applications, it is desirable to be able to describe the outcome of an experiment with a single number rather than verbally. In some cases, this is done automatically. That is, in rolling a die, the elementary outcomes are already represented by the numbers 1, 2, 3, 4, 5, 6. Of course, in any case with a finite number of elementary outcomes, it is possible to number the elements of the sample space $S = \{s_1, s_2, \ldots, s_n\}$ and use the subscript as the numerical description of an outcome. However, whatever means is used, a rule of assignment is defined which assigns to each element in the sample space a real number. Thus, we have a function whose domain is the sample space and whose range is some subset of the real numbers. Such a function is called a *random variable.*

Unfortunately, the standard terminology for random variables is different from the functional notation with which the reader is familiar. We will use s to symbolize a typical point in the sample space, X (or Y or $Z\ldots$) as the function and x (or y or $z\ldots$) as the values of function. That is, we will write $x = X(s)$. Of course, we could use different notation, but only at the price of making all other books on the subject unintelligible to students who have learned from this one.

Some typical examples of random variables follow.

Example 1 A coin is flipped. We assign 1 to heads and 0 to tails. That is, $X(H) = 1$, $X(T) = 0$. This could be said to be $X(s) =$ number of heads.

Example 2 A coin is flipped four times. Let $X(s) =$ number of heads. Now the range of $X = \{0, 1, 2, 3, 4\}$. For example, $X(H, H, T, H) = 3$.

Example 3 Two dice (one red, one green) are rolled. Let $Y(s) =$ sum of the two dice. Now, range of $Y = \{2, 3, 4, \ldots, 12\}$. For example, $Y(R2, G6) = 8$.

Notice that the outcome of a random variable is random, and one may assign probabilities to it in the natural way. That is, $P(X = x) = P\{s : X(s) = x\}$. For example, in Example 1 $P(X = 1) = 1/2$.

In Example 2, suppose we wanted $P(X = 3)$. That is, the probability of getting three heads in four flips of a fair coin. $P(X = 3) = P(HHHT, HHTH, HTHH, THHH)$. That is, $\{s : X(s) = 3\}$ contains four of the elementary outcomes. The reader may check that there are sixteen all together. Hence, $P(X = 3) = 4/16 = 1/4$.

In order to save writing, we shall use the symbol $p(x)$ for $P(X = x)$. This is called the *probability function.* Thus, associated with each random variable X, there is a probability function, $p(x) = P(X = x)$. Sometimes we display this information in tabular form. Let us consider Example 3 in some detail. In this case, the range of Y is $\{2, 3, 4, 5, \ldots, 12\}$. The probability function $p(y)$ could be displayed as follows:

y	2	3	4	5	6	7	8	9	10	11	12
$p(y)$	1/36	2/36	3/36	4/36	5/36	6/36	5/36	4/36	3/36	2/36	1/36

Note that the sum of the values of $p(y)$ over all y in the range of Y must add up to 1. It is possible to express $p(y)$ as

$$p(y) = \frac{6 - |7 - y|}{36} \qquad y = 2, 3, \ldots, 12$$

When it is not too complicated, we shall try to write $p(x)$ as a formula valid for all suitable values of x. Otherwise we shall use a table.

In addition, any function of a random variable will be a random variable. In the preceding, let $X = (Y - 6)^2$. Range of X is $\{0, 1, 4, 9, 16, 25, 36\}$, where $P(X = 4) = P(Y = 4 \text{ or } Y = 8) = P(Y = 4) + P(Y = 8) = 8/36$, for example. The probability function is

x	0	1	4	9	16	25	36
$p(x)$	5/36	10/36	8/36	6/36	4/36	2/36	1/36

EXERCISES

1. Write the probability function for X in Example 1.
2. Write the probability function for X in Example 2.
3. For the example in Exercise 2, find $P(X \geq 3)$ and $P(1 \leq X < 4)$.
4. Referring to Adams, Bailey and Calhoun in Section 4.2, let $Z =$ number of winners in the election. Find the probability function for Z.
5. Referring to Exercise 4, $Y = 3 - Z =$ number of losers, find the probability function for Y.
6. Referring to Exercise 2, write the probability function for $Y = (X - 2)^2$.
7. A box contains three white balls numbered 1, 2, 3 and three black balls numbered 1, 2, 3. Two balls are drawn at random (simultaneously) from the box. Write all the points in the sample space. Let $X =$ largest number drawn. Find the probability function for X. Let $Y =$ the number on the highest numbered white ball drawn. Find the probability function for Y.
8. Referring to Exercise 7, let $Z =$ sum of all white ball numbers minus the sum of all black ball numbers drawn. Find the probability function for Z.

9. Referring to Exercise 7, let W = difference between the numbers drawn. Find the probability function for W.

10. Referring to Exercises 7, 8, 9, find the probability functions for $(X - 1)^2$, $(Y - 1)^2$, $(Z - 1)^2$ and $(W - 1)^2$.

4.5 Expected Values and Vector Multiplication

We are now in a position to formalize the concept with which this chapter began. That is, what is the expected outcome of a random experiment? Suppose that a random variable X has k possible values, $x_1, x_2, \ldots, x_k$ with $p(x_1) = p_1, p(x_2) = p_2, \ldots, p(x_k) = p_k$. Suppose further that the experiment is performed N times and that x_1 is the outcome n_1 times, x_2 is the outcome n_2 times and so on. The average value of the outcomes will be given by the sum of all the outcomes, $n_1x_1 + n_2x_2 + \cdots + n_kx_k$ divided by N, the number of trials. Hence,

$$\begin{aligned}\text{Average} &= [n_1x_1 + n_2x_2 + \cdots + n_kx_k] \div N \\ &= x_1\frac{n_1}{N} + x_2\frac{n_2}{N} + \cdots + x_k\frac{n_k}{N} \\ &= x_1f_1 + x_2f_2 + \cdots + x_kf_k\end{aligned}$$

where $f_1, f_2, \ldots, f_k$ are the frequencies of occurrence of the various values. Now these frequencies should be approximately equal to the corresponding probabilities. Hence, we expect that over a large number of trials the average value of X will be $x_1p_1 + x_2p_2 + \cdots + x_kp_k$. We will symbolize this value by $E(X)$, called the *expected value of* X, that is, $E(X) = x_1p_1 + x_2p_2 + \cdots + x_kp_k$.

For example, let X = the outcome in rolling one die. The possible values of X are (1, 2, 3, 4, 5, 6). The corresponding probabilities are all 1/6. Hence, $E(X) = 1(1/6) + 2(1/6) + 3(1/6) + 4(1/6) + 5(1/6) + 6(1/6) = 21/6 = 3\frac{1}{2}$. Does this mean that when we roll a die we can expect to roll a $3\frac{1}{2}$? Of course not! In fact, we can never roll $3\frac{1}{2}$. But in a large number of rolls, the average of all the rolls should be close to $3\frac{1}{2}$. Consider the 200 trial run given on page 67. We have 27 ones totaling 27, 37 twos totaling 74, 39 threes totaling 117, etc. The grand total is 709. Dividing by 200, the total number of rolls, we find that the average value is 3.545, which is fairly close to 3.5.

Alternatively, we could have computed the average by taking $x_1f_1 + x_2f_2 + \cdots + x_6f_6 = 1(.135) + 2(.185) + 3(.195) + 4(.140) + 5(.175) + 6(.170)$, but it was easier to work with whole numbers.

Since expressions of the previous kind occur frequently in probability, as well as in other branches of mathematics, it is convenient to introduce some notation which will save considerable writing. We will use the symbol **x** (x boldface) to stand for the matrix

$$(x_1, x_2, \ldots, x_k)$$

which has only one row and k columns. This is called a "$1 \times k$ matrix" or "row matrix" or "row vector." We will symbolize the "$k \times 1$ matrix" (column matrix or column vector) of probabilities by

$$\mathbf{p} = \begin{pmatrix} p_1 \\ p_2 \\ \vdots \\ p_k \end{pmatrix}$$

Finally we will write

$$\mathbf{xp} = x_1p_1 + x_2p_2 + \cdots + x_kp_k$$

as the product of a row vector and column vector. You may wonder why we should write the first as a row and the second as a column and not simply both as rows. After all, it is easier to write numbers in a row and it takes less space on the page. Nevertheless, this is the convention which is used universally and it has its advantages later on. So, let us simply accept it and write

$$E(X) = \mathbf{xp}$$

where **x** is the row vector of possible values and **p** is the column vector of the corresponding probabilities.

For the actual average value of X in the 200-trial experiment, we have

$$\text{Average} = \mathbf{xf}$$
$$= (1, 2, 3, 4, 5, 6)\begin{pmatrix} .135 \\ .185 \\ .195 \\ .140 \\ .175 \\ .170 \end{pmatrix}$$

where, of course, **f** is the "frequency vector."

Let us try some other examples.

Example 1 A coin is flipped three times. Let X = number of heads. Here

$$\mathbf{x} = (0, 1, 2, 3) \qquad \mathbf{p} = \begin{pmatrix} 1/8 \\ 3/8 \\ 3/8 \\ 1/8 \end{pmatrix}$$

$$\begin{aligned} E(X) = \mathbf{xp} &= (0, 1, 2, 3)\begin{pmatrix} 1/8 \\ 3/8 \\ 3/8 \\ 1/8 \end{pmatrix} \\ &= 0(1/8) + 1(3/8) + 2(3/8) + 3(1/8) = 12/8 = 1\tfrac{1}{2} \end{aligned}$$

So we should average $1\frac{1}{2}$ heads out of three flips.

Example 2 Two dice are rolled. Y = sum of the faces.

$$\mathbf{y} = (2, 3, 4, \ldots, 12) \qquad \mathbf{p} = \begin{pmatrix} 1/36 \\ 2/36 \\ \vdots \\ 1/36 \end{pmatrix}$$

$$E(Y) = \mathbf{yp} = 7$$

The reader should check this one out. Also note that the expected sum on two dice is simply twice the expected value of a single die!

Interestingly enough, in Example 1, $E(X)$ is a number which X can never equal, and in Example 2, $E(Y)$ is the most likely value possible. This simply indicates that the expected value may or may not be a value which will come up frequently.

Since functions of random variables will be random variables themselves, it is possible to discuss the expected value of $Y = g(X)$. We may represent this as $E[g(X)]$, or simply $E(Y)$. The computation of this quantity may be carried out by finding the probability function for Y, $p_Y(y)$ and computing the expected value in the usual way or by simply finding the product

$$(g(x_1), g(x_2), \ldots, g(x_k))\begin{pmatrix} p_1 \\ p_2 \\ \vdots \\ p_k \end{pmatrix}$$

For example, in rolling a die, let $Y = (X - 3)^2$ where X is the number which appears. We now have

$$\mathbf{y} = (0, 1, 4, 9)$$

with corresponding probability vector

$$\mathbf{p}_Y = \begin{pmatrix} 1/6 \\ 1/3 \\ 1/3 \\ 1/6 \end{pmatrix}$$

and

$$\begin{aligned} E(Y) &= (0, 1, 4, 9)\begin{pmatrix} 1/6 \\ 1/3 \\ 1/3 \\ 1/6 \end{pmatrix} \\ &= 0(1/6) + 1(1/3) + 4(1/3) + 9(1/6) = 19/6 = 3\tfrac{1}{6} \end{aligned}$$

Alternatively

$$\begin{aligned} E(g(X)) &= ((-2)^2, (-1)^2, 0^2, 1^2, 2^2, 3^2)\begin{pmatrix} 1/6 \\ 1/6 \\ 1/6 \\ 1/6 \\ 1/6 \\ 1/6 \end{pmatrix} \\ &= 4(1/6) + 1(1/6) + 0(1/6) + 1(1/6) + 4(1/6) + 9(1/6) \\ &= 19/6 = 3\tfrac{1}{6} \end{aligned}$$

Usually we shall use the latter method to avoid having to compute a new probability function.

Let us realize that vector multiplication is not limited to probabilistic applications. As a simple example, we might have the following information: A store sells milk, butter and eggs. Milk costs 30¢/quart, butter costs 80¢/pound and eggs cost 50¢/dozen. An order comes in for four quarts of milk, three pounds of butter and two dozen eggs. If we let

$$\mathbf{R} = (4, 3, 2)$$

be the vector of amounts ordered in each category and

$$\mathbf{P} = \begin{pmatrix} .30 \\ .80 \\ .50 \end{pmatrix}$$

be the vector of price per unit item, then

$$\mathbf{RP} = (4,\ 3,\ 2)\begin{pmatrix} .30 \\ .80 \\ .50 \end{pmatrix} = 1.20 + 2.40 + 1.00 = \$4.60$$

gives the total amount of the sale. The nice thing about this is that we may say, "The number of items times the price per item equals the total cost," which is true even though the expressions **R** and **P** are not simply numbers but arrays of numbers.

EXERCISES

1. Calculate the expected number of heads in one flip of a fair coin, two flips, three flips, four flips. Can you guess a general formula?

2–8. Calculate the expected values of the random variables in Exercises 4–10, Section 4.4.

9. In a state lottery, tickets cost 50¢. Out of every million tickets, one is worth \$50,000, ten are worth \$5,000, 100 are worth \$500, 1,000 are worth \$50 and the rest are worth nothing. Find the expected value of a ticket.

10. In roulette, there are eighteen red numbers, eighteen black numbers and two green numbers. If you bet on red, you win \$1.00 if red comes up; you lose \$1.00 otherwise. All numbers are equally likely to come up. Find the expected value of your bet.

11. Calculate:

(a) $(1, -1, 2)\begin{pmatrix} 0 \\ 1 \\ 7 \end{pmatrix}$

(b) $(1/2, 0, 3, 1)\begin{pmatrix} 2 \\ 5 \\ 1 \\ -6 \end{pmatrix}$

12. Solve for x:

(a) $(x, 1, 2)\begin{pmatrix} 1 \\ x \\ -1 \end{pmatrix} = 6$

(b) $(x, 1, -1)\begin{pmatrix} x \\ 2x \\ -2 \end{pmatrix} = 5$

(c) $(x, -1)\begin{pmatrix} 1 \\ x \end{pmatrix} = 2$

(d) $(1, x, 1)\begin{pmatrix} 2 \\ 1 \\ -x \end{pmatrix} = 2$

13. Solve for x and y:

(a) $(x, y)\begin{pmatrix} 1 \\ -1 \end{pmatrix} = 3$ and $(x, y)\begin{pmatrix} 3 \\ 3 \end{pmatrix} = 7$

(b) $(2, 7)\begin{pmatrix} x \\ y \end{pmatrix} = 4$ and $(1, 3)\begin{pmatrix} x \\ y \end{pmatrix} = 6$

14. Let (x, y) be any point in the plane. Can you see what

$$(x, y)\begin{pmatrix} x \\ y \end{pmatrix}$$

represents geometrically?

4.6 Properties of Expectation

In this section, we shall consider some of the properties of the expected value. Since we have chosen to express the idea in terms of product of vectors, it is convenient to first discuss some of the algebra of vectors. For example, let $\mathbf{x}$ and $\mathbf{y}$ be row vectors, each having k entries, that is, $\mathbf{x} = (x_1, x_2, \ldots, x_k)$ and $\mathbf{y} = (y_1, y_2, \ldots, y_k)$. Let $\mathbf{p}$ be a k entry column vector. Now

$$\begin{aligned} \mathbf{xp} + \mathbf{yp} &= (x_1p_1 + x_2p_2 + \cdots + x_kp_k) + (y_1p_1 + y_2p_2 + \cdots + y_kp_k) \\ &= (x_1 + y_1)p_1 + (x_2 + y_2)p_2 + \cdots + (x_k + y_k)p_k \end{aligned}$$

Thus if we define

$$\mathbf{x} + \mathbf{y} = (x_1 + y_1, x_2 + y_2, \ldots, x_k + y_k)$$

we have

$$\mathbf{xp} + \mathbf{yp} = (\mathbf{x} + \mathbf{y})\,\mathbf{p}$$

In other words, we have defined the sum of two vectors to be the vector each of whose entries (called *components*) is the sum of corresponding components of the original vectors. We would also extend this definition to the sum of two column vectors. As a result of this definition, the following properties are easy to see:

1. The sum of two vectors is defined only if both are rows (or columns) having the same number of components. You cannot add a row vector to a column vector!
2. $\mathbf{x} + \mathbf{y} = \mathbf{y} + \mathbf{x}$ (commutative law)
3. $(\mathbf{x} + \mathbf{y}) + \mathbf{z} = \mathbf{x} + (\mathbf{y} + \mathbf{z})$ (associative law)
4. (a) $(\mathbf{x} + \mathbf{y})\,\mathbf{p} = \mathbf{xp} + \mathbf{yp}$
   (b) $\mathbf{p}(\mathbf{x} + \mathbf{y}) = \mathbf{px} + \mathbf{py}$ (distributive laws)

NOTE: In 4(a), **x** and **y** must be row vectors and **p** a column vector, while in 4(b) **x** and **y** must be column vectors and **p** must be a row vector. So, for example,

$$(1, 3, 5)\left[\begin{pmatrix} 2 \\ 0 \\ 6 \end{pmatrix} + \begin{pmatrix} -1 \\ 2 \\ 7 \end{pmatrix}\right] = (1, 3, 5)\begin{pmatrix} 1 \\ 2 \\ 13 \end{pmatrix} = 1 + 6 + 65 = 72$$

Alternatively,

$$\begin{aligned}(1, 3, 5)\left[\begin{pmatrix} 2 \\ 0 \\ 6 \end{pmatrix} + \begin{pmatrix} -1 \\ 2 \\ 7 \end{pmatrix}\right] &= (1, 3, 5)\begin{pmatrix} 2 \\ 0 \\ 6 \end{pmatrix} + (1, 3, 5)\begin{pmatrix} -1 \\ 2 \\ 7 \end{pmatrix} \\ &= (2 + 0 + 30) + (-1 + 6 + 35) = 32 + 40 = 72\end{aligned}$$

Since the sum of two vectors has been defined, it is now natural to wish to define the difference of two vectors. The simplest way to do this is to say

$$(x_1, x_2, \ldots, x_k) - (y_1, y_2, \ldots, y_k) = (x_1 - y_1, x_2 - y_2, \ldots, x_k - y_k)$$

This leads one to the result

$$\mathbf{x} - \mathbf{x} = \mathbf{0}$$

where the zero vector $\mathbf{0}$ has all components zero. Also, $\mathbf{0} - \mathbf{x} = -\mathbf{x}$, where $-\mathbf{x}$ is the vector $(-x_1, -x_2, \ldots, -x_k)$. Finally, of course, one would want to say $\mathbf{x} + \mathbf{x} = 2\mathbf{x}$, which leads us to defining the vector $c\mathbf{x}$ (where c is any number) by the rule

$$c\mathbf{x} = (cx_1, cx_2, \ldots, cx_k)$$

This definition has the added advantage of yielding

$$-\mathbf{x} = (-1)\mathbf{x} \qquad \mathbf{0} = 0\mathbf{x} \quad \text{and} \quad (c\mathbf{x})\mathbf{y} = c(\mathbf{xy})$$

Naturally, all these operations can be just as easily carried out with column vectors.

Let us see how this "vector algebra" can be applied to problems involving expected value. Suppose X is a random variable, $Y = f(X)$ and $Z = g(X)$. Suppose we wish to find $E[af(X) + bg(X)]$. Let

$$\mathbf{y} = [f(x_1), f(x_2), \ldots, f(x_k)]$$

and

$$\mathbf{z} = [g(x_1), g(x_2), \ldots, g(x_k)]$$

Now

$$\begin{aligned} E[af(X) + bg(X)] &= (a\mathbf{y} + b\mathbf{z})\mathbf{p} \\ &= (a\mathbf{y})\mathbf{p} + (b\mathbf{z})\mathbf{p} = a(\mathbf{yp}) + b(\mathbf{zp}) \\ &= aE(Y) + bE(Z) = aE[f(X)] + bE[g(X)] \end{aligned}$$

For example, suppose a bag contains four balls numbered 1, 2, 3, 4. A ball is chosen at random and X is the number of the ball. Now

$$\mathbf{x} = (1, 2, 3, 4) \qquad \mathbf{p} = \begin{pmatrix} 1/4 \\ 1/4 \\ 1/4 \\ 1/4 \end{pmatrix}$$

$$E(X) = (1, 2, 3, 4)\begin{pmatrix} 1/4 \\ 1/4 \\ 1/4 \\ 1/4 \end{pmatrix} = 1/4 + 1/2 + 3/4 + 1 = 2\tfrac{1}{2}$$

Suppose one wished to know $E(X^2 + 2X)$. We would write

$$\begin{aligned} E(X^2 + 2X) &= E(X^2) + 2E(X) \\ &= (1, 4, 9, 16)\begin{pmatrix} 1/4 \\ 1/4 \\ 1/4 \\ 1/4 \end{pmatrix} + 2(2\tfrac{1}{2}) \\ &= 1/4 + 1 + 9/4 + 4 + 5 = 12\tfrac{1}{2} \end{aligned}$$

What about $E[(X+1)^2]$? This would be

$$E[X^2 + 2X + 1] = 12\tfrac{1}{2} + E(1)$$

What is $E(1)$? We could work indirectly, letting

$$(X+1)^2 = Y \qquad \mathbf{y} = (4, 9, 16, 25)$$

$$E(Y) = (4, 9, 16, 25)\begin{pmatrix} 1/4 \\ 1/4 \\ 1/4 \\ 1/4 \end{pmatrix} = 1 + 9/4 + 4 + 25/4 = 13\tfrac{1}{2}$$

Therefore, $13\tfrac{1}{2} = 12\tfrac{1}{2} + E(1)$ and $E(1) = 1$. It would be nice if this were always the case, and it is. In fact, let c be any constant. Now for all values of X, c is equal to c; hence

$$\begin{aligned} E(c) &= cp_1 + cp_2 + \cdots + cp_k \\ &= c(p_1 + p_2 + \cdots + p_k) = c(1) = c \end{aligned}$$

since the sum of the probabilities is always one.

As a last example, let us consider the following example: A biased coin lands heads with probability 0.5, tails with probability 0.3 and edge with probability 0.2. Suppose we agree to the following game: If heads comes up, you pay me \$2; if edge comes up, you pay me \$1; and if tails comes up, you pay me nothing. Letting X be the amount which you pay, then $\mathbf{x} = (2, 1, 0)$ and

$$E(X) = (2, 1, 0)\begin{pmatrix} .5 \\ .2 \\ .3 \end{pmatrix} = 1.2$$

If I pay you \$1.20 to persuade you to play, then the payoffs would be $Y = X - 1.2$ and $E[X - 1.2] = E(X) - 1.2 = 1.2 - 1.2 = 0$. Thus we have a fair game. That is, on the average we both expect to break even.

EXERCISES

1. Two fair dice are rolled. Let X be the sum of the uppermost faces. Find $E[(X-7)^2]$ directly and by expanding $(X-7)^2$ and applying the properties of expectation.
2. Repeat Exercise 4, Section 4.5, by expanding $(X-2)^2$ and using the properties of expectation.

3. Repeat Exercise 8, Section 4.5, by expanding and using the properties of expectation.

4. Given $E(X) = 1$, $E(X^2) = 3$ and $E(X^3) = 2$, find $E[(X - 1)^3]$.

5. Given $E[(X + 3)^2] = 19$ and $E(X) = -1$, find $E(X^2)$.

6. Given $E[(X + 3)^2] = 19$ and $E[(X - 3)^2] = 3$, find $E(X^2)$ and $E(X)$.

7. Evaluate:

$$(-1, 3, 2)\left[3\begin{pmatrix}2\\-1\\1\end{pmatrix} - 2\begin{pmatrix}1\\-1\\3\end{pmatrix}\right]$$

8. Evaluate:

$$(1, -1, 2) + 2(-1, 3, 7) - 3(1, 0, 5)$$

9. Solve for x:

$$[(1, x, 2) + (2, 1, 3)]\begin{pmatrix}2\\-1\\4\end{pmatrix} = 7$$

10. Solve for x:

$$[(x, 2) - (3, x)]\begin{pmatrix}x\\-x\end{pmatrix} = x^2$$

11. Solve for a, b, c:

$$a(1, 2, 0) + b(-1, 1, 2) + c(-2, 1, 3) = (1, 6, -3)$$

12. Show that if $\mathbf{0}$ is a zero vector of suitable size and shape, then $\mathbf{x} \cdot \mathbf{0} = 0$ for any vector $\mathbf{x}$.

13. Find x and y, if $(x, y)\begin{pmatrix}x\\y\end{pmatrix} = 1$ and $(x, y)\begin{pmatrix}1\\-2\end{pmatrix} = 0$.

14. From Exercise 13, you can see that if the product of two vectors is zero, this does not imply that one of the two vectors is zero. Now, find all vectors (x, y, z) so that

$$(x, y, z)\begin{pmatrix}1\\-1\\2\end{pmatrix} = 0$$

and

$$(x, y, z)\begin{pmatrix}-1\\1\\0\end{pmatrix} = 0$$

4.7 Variance

As we have seen, the expected value of a random variable is the weighted average of the points in its range. The probabilities play the roles of the weights and hence the more likely a value is, the more heavily it is counted in finding the average. For this reason, $E(X)$ is usually called the *mean* of X and is thought of as the center about which observations of X will distribute themselves. Of course, the mean alone does not really give that much information. Consider, for example, the following probability functions:

x	$p_1(x)$	$p_2(x)$	$p_3(x)$
1	.2	.4	.05
2	.2	.1	.1
3	.2	0	.7
4	.2	.1	.1
5	.2	.4	.05

For each of the probability functions $p_1(x)$, $p_2(x)$ and $p_3(x)$ the mean is 3. In each case, the probabilities are symmetrically distributed about the mean. Nevertheless, one would expect quite different behavior if observations were made of random variables having these probability functions. For $p_1(x)$, one would expect about equal numbers of observations of all five of the values 1, 2, 3, 4, 5. In the second case, $p_2(x)$, most of the observations should be ones and fives, far from the mean. In the third case, most observations should be right at the mean. In this section, we shall describe a measurement called the *variance* which describes how spread out the observations are expected to be.

To fix the ideas, let us denote the expected value or mean of the random variable X by μ (the Greek letter mu). The mean is in some sense the central point about which observations of X will distribute. For short, we say "μ is the center of the distribution." We are interested in measuring the spread of the distribution. Since $X - \mu$ is the difference between X and its mean, one might wish to find the average or expected value of this quantity. However, $E[X - \mu] = E(X) - \mu = \mu - \mu = 0$. So this gives no information. A better choice might be $E[|X - \mu|]$. The absolute value eliminates negative numbers and hence avoids having the average come out zero. However, absolute values are algebraically annoying. So instead we use $E[(X - \mu)^2]$. This measurement is called the *variance* of X. It is abbreviated $\text{Var}(X)$. $\text{Var}(X)$ has the drawback that it involves square units. That is, if X is measured in feet, $\text{Var}(X)$ would be in square feet.

Therefore, many people prefer to use $\sqrt{\text{Var}(X)}$. This expression is called the "standard deviation of X." It is usually denoted by σ (the Greek letter sigma). Accordingly, the variance is denoted by σ^2. Summarizing, we have

$$\mu = E(X) \tag{1}$$

$$\sigma^2 = \text{Var}(X) = E[(X - \mu)^2] \tag{2}$$

$$\sigma = \text{SD}(X) = \sqrt{\text{Var}(X)} \tag{3}$$

Let us now contrast the three distributions given at the beginning of this section. For the distribution $p_1(x)$ we have,

$$\begin{aligned}\text{Var}(X) &= E[(X - \mu)^2] = E[(X - 3)^2] \\ &= (-2)^2(0.2) + (-1)^2(0.2) + 0^2(0.2) + (1)^2(0.2) + (2)^2(0.2) = 2\end{aligned}$$

and

$$\sigma = \sqrt{2}$$

For the distribution $p_2(x)$,

$$\begin{aligned}\text{Var}(X) &= (-2)^2(0.4) + (-1)^2(0.1) + (1)^2(0.1) + (2)^2(0.4) \\ &= 1.6 + 0.1 + 0.1 + 1.6 = 3.4\end{aligned}$$

and

$$\sigma = \sqrt{3.4}$$

Finally, for $p_3(x)$,

$$\begin{aligned}\text{Var}(X) &= (-2)^2(0.05) + (-1)^2(0.1) + 0^2(0.7) + (1)^2(0.1) + (2)^2(0.05) \\ &= 0.2 + 0.1 + 0 + 0.1 + 0.2 = 0.6\end{aligned}$$

and

$$\sigma = \sqrt{0.6}$$

It is easy to see that the more spread out distribution, the larger the variance. Let us consider another example. Suppose we have the following distribution:

x	1	2	3	4	5	6	7	8
$p_4(x)$	.3	.2	.1	.05	.05	.1	.1	.1

In this case, $E(X) = 0.3 + 0.4 + 0.3 + 0.2 + 0.25 + 0.6 + 0.7 + 0.8 = 3.55$. To compute $E[(X - 3.55)^2]$ now would be quite a nuisance. However, there is a very nice little identity which can ease the task. Remembering that μ is a constant, we can say

$$\begin{aligned} \text{Var}(X) &= E[(X - \mu)^2] = E[X^2 - 2\mu X + \mu^2] \\ &= E(X^2) - 2\mu E(X) + E(\mu^2) \\ &= E(X^2) - 2\mu\mu + \mu^2 \\ &= E(X^2) - 2\mu^2 + \mu^2 = E(X^2) - \mu^2 \end{aligned}$$

Thus, in general, $\text{Var}(X) = E(X^2) - \mu^2$. Notice, $E(X^2) \neq [E(X)]^2$. Now, in our last example,

$$\begin{aligned} E(X^2) &= 1^2(0.3) + 2^2(0.2) + 3^2(0.1) + 4^2(0.05) + 5^2(0.05) + 6^2(0.1) \\ &\quad + 7^2(0.1) + 8^2(0.1) \\ &= 18.95 \end{aligned}$$

Hence,

$$\begin{aligned} \text{Var}(X) &= 18.95 - \mu^2 = 18.95 - (3.55)^2 \\ &= 18.95 - 12.6025 = 6.3475 \end{aligned}$$

With the aid of this last identity, we have a very nice way to compute the variance. We feel that the variance should measure the spread of the distribution. The question is, in what sense? A very famous inequality discovered by Chebyschev provides a partial answer. In words it says that the probability of having an observation more than k standard deviations away from μ is less than $1/k^2$. That is, the probability that an observation is more than two standard deviations from μ is less than $1/4$. An observation more than three standard deviations from μ occurs less than $1/9$ of the time and so on.

Another way of saying this is to say that the probability of an observation falling between $\mu - k\sigma$ and $\mu + k\sigma$ is at least $1 - 1/k^2$. Or in symbols,

$$P[\mu - k\sigma \leq X \leq \mu + k\sigma] \geq 1 - 1/k^2$$

Let us reconsider the three examples we have for the case $k = 2$. For $p_1(x)$, $\sigma = \sqrt{2} \approx 1.4$ and $\mu = 3$. Chebyschev's inequality says

$$\begin{aligned} &P[3 - 2.8 \leq X \leq 3 + 2.8] \geq 1 - 1/4 \\ &P[0.2 \leq X \leq 5.8] \geq 3/4 \end{aligned}$$

In fact, X is always between 1 and 5. Hence,

$$P[0.2 \leq X \leq 5.8] = 1 \geq 3/4$$

The case $p_2(x)$ gives similar results. For the case $p_3(x)$, $\sigma = \sqrt{0.6} \approx 0.77$, $\mu = 3$. The inequality gives

$$P[3 - 1.54 \leq X \leq 3 + 1.54] \geq 3/4$$
$$P[1.46 \leq X \leq 4.54] \geq 3/4$$

Now,

$$P[1.46 \leq X \leq 4.54] = P[X = 2, 3, 4] = 0.9 \geq 3/4$$

as predicted.

Finally, for $p_4(x)$, $\sigma = \sqrt{6.3475} \approx 2.52$, $\mu = 3.55$. The case $k = 2$ would again yield a true probability of 1, so let us try $k = 1.5$. Chebyschev's inequality states

$$P[3.55 - (1.5)(2.52) \leq X \leq 3.55 + (1.5)(2.52)] \geq 1 - 1/(1.5)^2$$
$$P[-0.23 \leq X \leq 7.33] \geq 0.56$$

In fact,

$$\begin{aligned} P[-0.23 \leq X \leq 7.33] &= P(X \neq 8) \\ &= 1 - P(X = 8) = 1 - 0.1 = 0.9 \geq 0.56 \end{aligned}$$

In every case, Chebyschev's inequality, although correct, does not give very precise values for the probability. Nevertheless, it gives an upper bound which is correct for *every* probability distribution. The proof of this remarkable result may be found in the appendix.

EXERCISES

1. Let X be the number which appears when a fair die is rolled. Find $\text{Var}(X)$.

2–8. Find the variances of the random variables whose expectations were computed in Exercises 2–8, Section 4.5.

Properties of Variance:

9. Let X be a random variable, with range $\{1, 2, \ldots, 8\}$ and let $p(x) = x/36$. Find $E(X)$ and $\text{Var}(X)$.

10. For X in Exercise 9, compare the Chebyschev probability upper bound with the true probability for

$$P[\mu - k\sigma \leq X \leq \mu + k\sigma] \quad \text{for } k = 1\tfrac{1}{2}, 2, 2\tfrac{1}{2}$$

11. A card is drawn from a standard deck. Let X be the number of the card where all pictures are 10 and aces are 11. Find the probability function for X, $E(X)$ and $\text{Var}(X)$. Now repeat Exercise 10 for this random variable.

12. If $\text{SD}(X)$ measures spread, then doubling X should double $\text{SD}(X)$. Show that if $Y = 2X$, $\text{Var}(Y) = 4\ \text{Var}(X)$ and hence $\text{SD}(Y) = 2\ \text{SD}(X)$.

13. Repeat Exercise 12 for $Y = -2X$.

14. Show that if c is any constant, $\text{Var}(cX) = c^2\text{Var}(X)$, and hence $\text{SD}(cX) = |c|\text{SD}(X)$.

15. Show that $\text{Var}(X + b) = \text{Var}(X)$ for any constant b. (This makes sense, since adding a constant to all values of X should not change the spread.)

4.8 Games of Strategy

We began this chapter with a discussion of a simple gambling game. Like many such games, this one involved pure chance with no strategic considerations. In this section, we shall discuss some competitive games which do not involve any chance. Of course, most games which are really of interest involve elements of both skill and chance. Real games of this type are far too complicated to be analyzed here.

We shall deal with a very special kind of game. To begin with, we shall always consider two-player games in which each player has only a finite number of possible plays. To be specific, let us say that Player I has m possible plays and Player II has n possible plays. Each player makes his play without the other knowing what he will do. When Player I picks his ith play and Player II picks his jth play, there results a "payoff" g_{ij} from Player II to Player I. Of course, g_{ij} may be negative, indicating that Player II is the winner. We may display the collection of outcomes as a matrix $\mathbf{G}$. The entries in $\mathbf{G}$ will be the numbers g_{ij} arranged as follows: Numbering the rows of $\mathbf{G}$, $1, 2, \ldots, m$ from top to bottom and the columns $1, 2, \ldots, n$ from left to right, then g_{ij} is the entry in Row i, Column j. For example,

$$\mathbf{G} = \begin{pmatrix} -1 & 2 & 3 \\ 1 & -4 & 5 \end{pmatrix}$$

In this case, $m = 2$, $n = 3$ and $\mathbf{G}$ is called a 2×3 matrix. The first player has two possible plays and the second player has three possible plays. If Player I plays his first possibility and Player II plays his third possibility, the payoff is g_{13}, the entry in Row 1 and Column 3, namely, 3. Thus, in this case Player II pays three units to Player I.

For our discussion of game strategies, it will be convenient to define the product of a matrix and a vector. To start, let $\mathbf{u} = (u_1, u_2, \ldots, u_m)$ be a row vector with m components. Let $\mathbf{G}$ be an $m \times n$ matrix:

$$\mathbf{G} = \begin{pmatrix} g_{11} & g_{12} & g_{13} & \cdots & g_{1n} \\ g_{21} & g_{22} & g_{23} & \cdots & g_{2n} \\ \vdots & & & & \vdots \\ g_{m1} & g_{m2} & g_{m3} & \cdots & g_{mn} \end{pmatrix}$$

The columns of $\mathbf{G}$ may be thought of as m-component column vectors, $\mathbf{C}_1, \mathbf{C}_2, \ldots, \mathbf{C}_n$ where

$$\mathbf{C}_i = \begin{pmatrix} g_{1i} \\ g_{2i} \\ \vdots \\ g_{mi} \end{pmatrix}$$

Hence we may define the product of $\mathbf{u}$ with each of the column vectors of $\mathbf{G}$. Let us define the product $\mathbf{uG}$ as the row vector whose components are $\mathbf{uC}_1, \mathbf{uC}_2, \ldots, \mathbf{uC}_n$. For example,

$$(3, 2, 1, -1)\begin{pmatrix} 2 & 0 \\ -1 & 1/2 \\ 2 & 1 \\ 1 & -4 \end{pmatrix} = (5, 6)$$

This is the product of a 4-component row vector with a 4×2 matrix. The result is a 2-component row vector with entries

$$(3, 2, 1, -1)\begin{pmatrix} 2 \\ -1 \\ 2 \\ 1 \end{pmatrix} = 5 \quad \text{and} \quad (3, 2, 1, -1)\begin{pmatrix} 0 \\ 1/2 \\ 1 \\ -4 \end{pmatrix} = 6$$

We also define the product $\mathbf{Gv}$ of an $m \times n$ matrix with an n-component column vector as a column vector whose entries are the products of the rows of $\mathbf{G}$ with $\mathbf{v}$. Thus,

$$\begin{pmatrix} 1 & -1 & 2 \\ -1 & 0 & 4 \end{pmatrix}\begin{pmatrix} 1 \\ -3 \\ 2 \end{pmatrix} = \begin{pmatrix} 8 \\ 7 \end{pmatrix}$$

If you find it difficult to remember, think of a row vector as a $1 \times m$ matrix and a column vector as an $n \times 1$ matrix. Then a $(\underline{1} \times m)$ times an $(m \times \underline{n})$ gives a $\underline{1} \times \underline{n}$, so in the first example we had a $(\underline{1} \times 4) \times (4 \times \underline{2})$ and ended with a $(\underline{1} \times \underline{2})$. Similarly, in the second example, we had a $(\underline{2} \times 3)$ times a $(3 \times \underline{1})$, yielding a $(\underline{2} \times \underline{1})$. In the next chapter we shall define the product of an $(\underline{m} \times n)$ matrix with an $(n \times \underline{p})$ matrix and the result will be $(\underline{m} \times \underline{p})$.

Let us now look at a simple (2×2) matrix game. Suppose each player has two cards; Player I has a red 1 and a black 6. Player II has a red 3 and a black 1. Each player plays a card. If the colors match, Player I wins the sum of the numbers played. Otherwise, Player II does. The payoff matrix for this game is

$$\mathbf{G} = \begin{pmatrix} 4 & -2 \\ -9 & 7 \end{pmatrix} \begin{matrix} R1 \\ B6 \end{matrix}$$
$$\quad R3 \quad B1$$

Now Player I would like to play $B6$ where he might win 7 units but he opens himself up to a loss of 9 units if his opponent guesses correctly. Similarly, $R1$ may offer both rewards and penalties. Therefore, Player I must avoid falling into any pattern of play in repetitions of the game for fear that his opponent may discover the pattern. Accordingly, Player I decides to play his cards randomly. To start, he may try playing each card 1/2 of the time. Thus, when his opponent plays $R3$, Player I expects to win

$$v_1 = (1/2)(4) + 1/2(-9) = -2\tfrac{1}{2}$$

When his opponent plays $B1$, Player I expects to win

$$v_2 = (1/2)(-2) + (1/2)(7) = +2\tfrac{1}{2}$$

We may write this in matrix notation as

$$(1/2, 1/2)\begin{pmatrix} 4 & -2 \\ -9 & 7 \end{pmatrix} = (-2\tfrac{1}{2}, 2\tfrac{1}{2})$$

If Player I chooses to play $R1$ with probability 1/3 and the other card with probability 2/3, we arrive at

$$(1/3, 2/3)\begin{pmatrix} 4 & -2 \\ -9 & 7 \end{pmatrix} = (-14/3, 12/3)$$

In this way, Player I increases his possible winnings but also his possible losses! Game theorists, being conservative, would prefer (1/2, 1/2) where the possible losses are minimized.

Now, in general, let $\mathbf{p} = (p, 1 - p)$ be the vector of probabilities with which Player I chooses his plays; p may be any number between 0 and 1. Now

$$(v_1, v_2) = (p, 1 - p)\begin{pmatrix} 4 & -2 \\ -9 & 7 \end{pmatrix}$$

$$(v_1, v_2) = (4p - 9 + 9p, -2p + 7 - 7p)$$

$$(v_1, v_2) = (13p - 9, 7 - 9p)$$

Thus,

$$v_1 = 13p - 9 \quad \text{and} \quad v_2 = 7 - 9p$$

Plotting these two equations, we get two intersecting straight lines (figure 3). They intersect at $p = 8/11$, $v_1 = v_2 = 5/11$. For values of $p < 8/11$, v_2 is greater than v_1 and for $p > 8/11$ the opposite is true. The conservative game theorist looks at the smaller possible value for each value of p, which represents the minimum possible expected winnings (which is sometimes negative). The graph of minimum possible winnings plotted versus p is indicated by the heavy broken line in figure 4. Clearly, Player I should pick p so as to hit the high point on this graph. But this is precisely the point of intersection of the two original lines. That is, the optimal value of p must be where $v_1 = v_2$. Hence we must solve

$$\mathbf{pG} = (v, v) \quad \text{where } \mathbf{p} = (p, 1 - p)$$

and $\mathbf{G}$ is the (2×2) payoff matrix. Of course, the resulting v is the expected winnings of Player I. This is called the *value of the game to Player I.*

Let us now look at the game from Player II's point of view. He also must play randomly and his strategy may be represented by the column vector

$$\mathbf{q} = \begin{pmatrix} q \\ 1 - q \end{pmatrix}$$

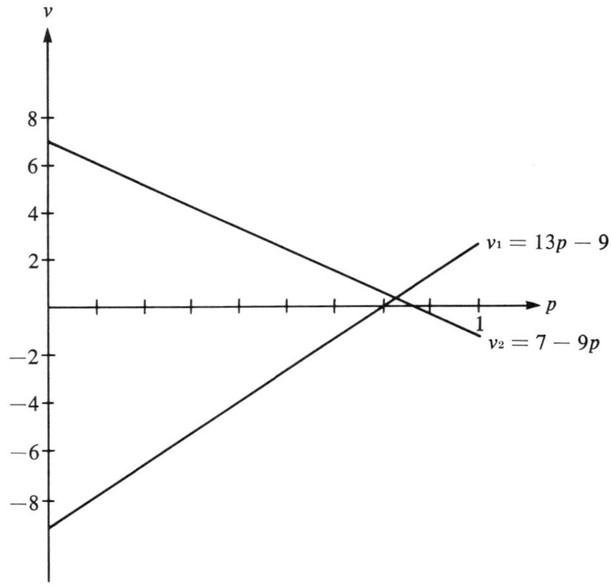

Figure 3

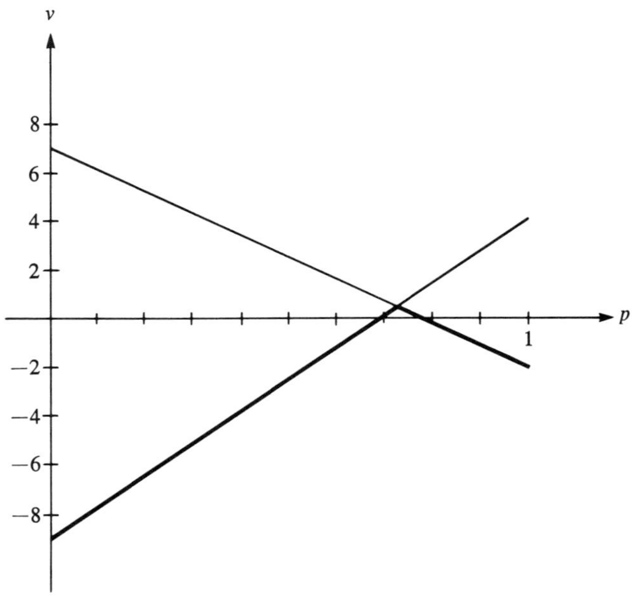

Figure 4

where q is the probability with which he plays the first column. The product

$$\mathbf{Gq} = \begin{pmatrix} 4 & -2 \\ -9 & 7 \end{pmatrix}\begin{pmatrix} q \\ 1-q \end{pmatrix} = \begin{pmatrix} 4q - 2 + 2q \\ -9q + 7 - 7q \end{pmatrix} = \begin{pmatrix} w_1 \\ w_2 \end{pmatrix}$$

yields the vector of expected values to Player II assuming that Player I plays Row 1 or Row 2. So we have

$$w_1 = 6q - 2$$
$$w_2 = -16q + 7$$

and plotting these two lines, we see that they intersect at $q = 9/22$, $w_1 = w_2 = 5/11$. Figure 5 shows these two lines. Remembering that the values of w represent the amount that Player II loses, we realize that he must always worry about the larger of the two possible values. The heavy line in figure 6 shows the graph of the larger value of w versus q. Surely Player II must aim for the low point on this graph, which is again the point of

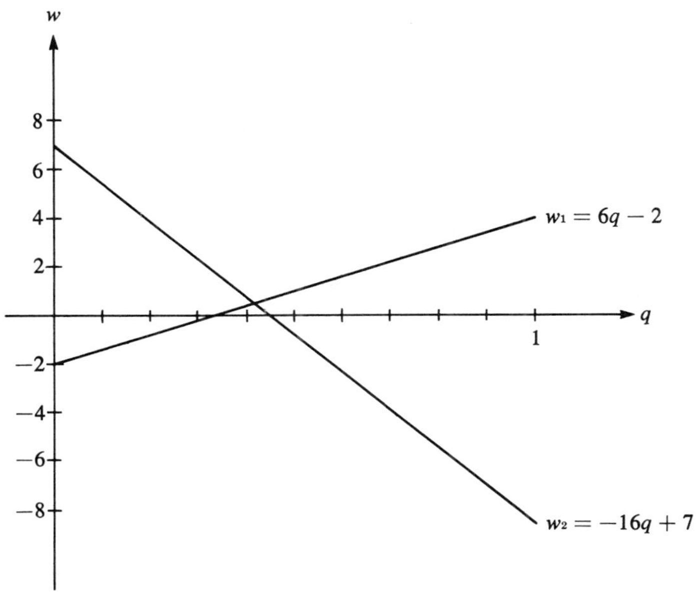

Figure 5

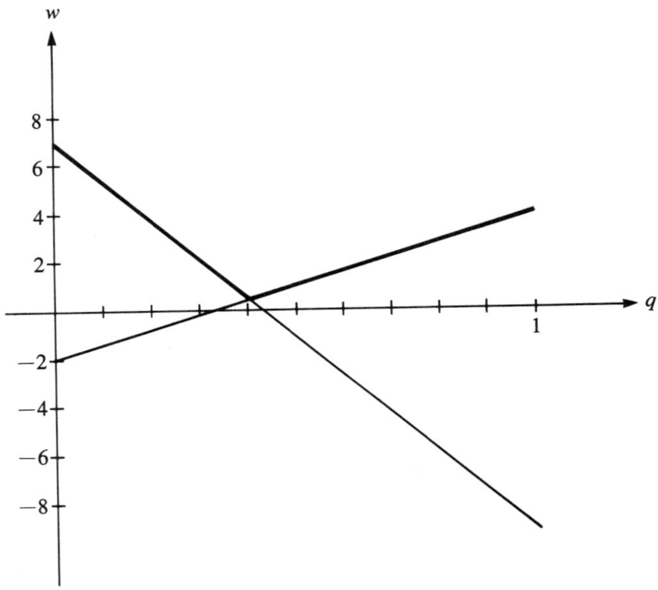

Figure 6

intersection of the two original lines. Hence, Player II's strategy must be arrived at by solving

$$\mathbf{Gq} = \begin{pmatrix} w \\ w \end{pmatrix}$$

Here we would call w the *value of the game to Player II.* By now the reader has observed that in this case $v = w = 5/11$. This is not a coincidence. It follows from the fact that the associative law holds for multiplication of matrices. That is,

$$(\mathbf{pG})\mathbf{q} = \mathbf{p}(\mathbf{Gq})$$

The reader is asked to prove this as Exercise 1 for the type of matrices encountered in this section. However, let us use it to prove the following theorem.

THEOREM 6 Let **G** be a (2×2) game matrix and suppose there are solutions to

$$\mathbf{pG} = (v, v) \quad \text{and} \quad \mathbf{Gq} = \begin{pmatrix} w \\ w \end{pmatrix}$$

Then $v = w$.

Proof By the associative law

$$(\mathbf{pG})\mathbf{q} = \mathbf{p}(\mathbf{Gq})$$

Therefore,

$$(v, v)\begin{pmatrix} q \\ 1 - q \end{pmatrix} = (p, 1 - p)\begin{pmatrix} w \\ w \end{pmatrix}$$

$$vq + v(1 - q) = pw + w(1 - p)$$

$$vq + v - vq = pw + w - pw$$

$$v = w$$

In summary, if **G** is a (2×2) game matrix, the optimal strategies of the two players may be sought by solving the equations

$$(p, 1 - p)\mathbf{G} = (v, v)$$

and

$$\mathbf{G}\begin{pmatrix} q \\ 1 - q \end{pmatrix} = \begin{pmatrix} v \\ v \end{pmatrix}$$

The vector $\mathbf{p} = (p, 1 - p)$ gives the first player's strategy, $\mathbf{q} = \begin{pmatrix} q \\ 1 - q \end{pmatrix}$ gives the second player's strategy and v gives the "value of the game." Finding **p**, **q** and v is called *solving the game*. Now it is possible that these equations do not submit to a solution with p and q between zero and one. In such cases another approach must be found, and we shall find it in the next section.

EXERCISES

1. Let $\mathbf{G} = \begin{pmatrix} a & b \\ c & d \end{pmatrix}$, $\mathbf{p} = (p_1, p_2)$ and $\mathbf{q} = \begin{pmatrix} q_1 \\ q_2 \end{pmatrix}$. Show that $(\mathbf{pG})\mathbf{q} = \mathbf{p}(\mathbf{Gq})$.

2. Verify the associative law $\mathbf{A}(\mathbf{BC}) = (\mathbf{AB})\mathbf{C}$ for

$$\mathbf{A} = (1, 3, -2) \qquad \mathbf{B} = \begin{pmatrix} -1 & 3 \\ 2 & 1 \\ -6 & 5 \end{pmatrix} \qquad \mathbf{C} = \begin{pmatrix} 1 \\ 2 \end{pmatrix}$$

3. Verify the distributive law,

$$(\mathbf{A} + \mathbf{B})\mathbf{C} = \mathbf{AC} + \mathbf{BC}$$

for

$$\mathbf{A} = (-1, 2, 1) \qquad \mathbf{B} = (4, -3, 0) \qquad \mathbf{C} = \begin{pmatrix} -1 & 4 & 1 & 2 \\ 6 & -1 & 5 & 0 \\ 2 & 0 & 1 & 3 \end{pmatrix}$$

4. Solve for x, y and z:

$$\begin{pmatrix} 3 & -1 & 2 \\ 0 & 1 & 3 \\ 4 & 2 & 1 \end{pmatrix} \begin{pmatrix} x \\ x \\ 3 \end{pmatrix} = \begin{pmatrix} 8 \\ y \\ z \end{pmatrix}$$

5. Solve for x, y and z:

$$\begin{pmatrix} -1 & 2 & 1 \\ 0 & 3 & 2 \\ 1 & 1 & -1 \end{pmatrix} \begin{pmatrix} x \\ y \\ z \end{pmatrix} = \begin{pmatrix} 2 \\ 1 \\ 7 \end{pmatrix}$$

6. Write the following system of equations in the form $\mathbf{AX} = \mathbf{B}$ where $\mathbf{A}$ is a (3×3) matrix, $\mathbf{X} = \begin{pmatrix} x \\ y \\ z \end{pmatrix}$ and $\mathbf{B}$ is a column vector:

$$\begin{aligned} 3x + 2y + z &= 2 \\ x - y - z &= 6 \\ y + z &= 5 \end{aligned}$$

Solve the equations.

7. Find the optimal strategies and the value of the game if

(a) $\mathbf{G} = \begin{pmatrix} -1 & 2 \\ 1 & -3 \end{pmatrix}$.

(b) $\mathbf{G} = \begin{pmatrix} 4 & 1 \\ 4 & 6 \end{pmatrix}$.

(c) $\mathbf{G} = \begin{pmatrix} -1 & -2 \\ -3 & 0 \end{pmatrix}$.

8. Suppose we play the same game as in the text but let the payoff be the product of the numbers. Write the payoff matrix and solve the game.

9. Joe Namath may call for a "pass play" or a "running play." The defense may prepare for a pass or a run. If the defense guesses wrong, Joe's team gains 20 yards on a pass or 10 yards on a run. If the defense guesses right, Joe's team gains nothing on a pass and 2 yards on a run. Write this as a matrix game and solve.

4.9 Row and Column Dominance: Saddle Points

Suppose we have a game payoff matrix

$$\mathbf{G} = \begin{pmatrix} 3 & 5 \\ 1 & 7 \end{pmatrix}$$

We try our standard method of attack, seeking the first player's strategy,

$$(p, 1 - p)\begin{pmatrix} 3 & 5 \\ 1 & 7 \end{pmatrix} = (v, v)$$

$$3p + 1 - p = v$$

$$5p + 7 - 7p = v$$

Equating the lefthand sides of these equations, we obtain

$$2p + 1 = 7 - 2p$$

$$4p = 6$$

$$p = 1\tfrac{1}{2}$$

Of course, $p = 1\frac{1}{2}$ is impossible since p, being a probability measurement, cannot exceed 1. Something must be wrong. Let us look at $\mathbf{G}$ more closely. From the first player's point of view it seems that a random strategy is necessary. However, let us take the second player's viewpoint. He sees that if Player I plays Row 1, he should play Column 1, since he loses three instead of five. If Player I plays Row 2, Player II should still play Column 1, since he loses one instead of seven. Thus, the second player will never play Column 2. Accordingly, we may as well strike Column 2 from consideration. Now Player I looks at the payoff matrix, seeing

$$\mathbf{G} = \begin{pmatrix} 3 & \not{5} \\ 1 & \not{7} \end{pmatrix}$$

Surely, he will now choose to always play Row 1 since this gains him three units instead of one. Thus, the first player's strategy is $\mathbf{p} = (1, 0)$, the second player's strategy is $\mathbf{q} = \begin{pmatrix} 1 \\ 0 \end{pmatrix}$ and the value of the game is three.

The idea here is that neither player will select a play if there is another one available which always gives better results. Formally we will say that $\mathbf{u} = (u_1, u_2, \ldots, u_k)$ dominates $\mathbf{v} = (v_1, v_2, \ldots, v_k)$ if $u_1 \geq v_1$, $u_2 \geq v_2, \ldots,$ $u_k \geq v_k$, with a similar definition for column vectors. Now, we have two rules for analysis of payoff matrices:

1. We may strike out any column which dominates any other column (since Player II will never choose it).
2. We may strike out any row which is dominated by any other (since Player I will never choose it).

Sometimes application of these two rules may simplify a complicated looking game. For example,

$$\mathbf{G} = \begin{pmatrix} 1 & 3 & 5 \\ 2 & 4 & 2 \\ -1 & 0 & -2 \\ -2 & 5 & -3 \end{pmatrix}$$

Inspecting **G**, we first check the rows. We see that Row 3 is dominated by Rows 1 and 2 and hence may be struck out, leaving

$$\mathbf{G} = \begin{pmatrix} 1 & 3 & 5 \\ 2 & 4 & 2 \\ \cancel{-1} & \cancel{0} & \cancel{-2} \\ -2 & 5 & -3 \end{pmatrix}$$

There are now no more row dominances. Let us look at the columns. We see that Column 2 dominates Column 1, leaving

$$\mathbf{G} = \begin{pmatrix} 1 & \cancel{3} & 5 \\ 2 & \cancel{4} & 2 \\ \cancel{-1} & \cancel{0} & \cancel{-2} \\ -2 & \cancel{5} & -3 \end{pmatrix}$$

In the reduced payoff matrix, Row 4 is now dominated by Rows 1 and 2; hence

$$\mathbf{G} = \begin{pmatrix} 1 & \cancel{3} & 5 \\ 2 & \cancel{4} & 2 \\ \cancel{-1} & \cancel{0} & \cancel{-2} \\ \cancel{-2} & \cancel{5} & \cancel{-3} \end{pmatrix}$$

Here Column 3 dominates Column 1

$$\mathbf{G} = \begin{pmatrix} 1 & \cancel{3} & \cancel{5} \\ 2 & \cancel{4} & \cancel{2} \\ \cancel{-1} & \cancel{0} & \cancel{-2} \\ \cancel{-2} & \cancel{5} & \cancel{-3} \end{pmatrix}$$

and finally Row 2 dominates Row 1, leaving

$$\mathbf{G} = \begin{pmatrix} 1 & 3 & 5 \\ 2 & 4 & 2 \\ -1 & 0 & -2 \\ -2 & 5 & -3 \end{pmatrix}$$

Thus, Player I's strategy is $\mathbf{p} = (0, 1, 0, 0)$. Player II's strategy is

$$\mathbf{q} = \begin{pmatrix} 1 \\ 0 \\ 0 \end{pmatrix} \quad \text{and} \quad v = 2$$

In both of these cases, we have what is known as a pure strategy. That is, each player is logically able to select only one play and no randomization is needed. There is another type of game which leads to a pure strategy solution. It is typified by the following example:

$$\mathbf{G} = \begin{pmatrix} 5 & -2 & -1 \\ 3 & 2 & 1 \\ -3 & 5 & -1 \end{pmatrix}$$

Player I makes a list of the worst that can happen to him if he plays pure strategies:

	Worst Possibility
Row 1	−2
Row 2	1
Row 3	−3

From a pessimistic point of view, all he can guarantee by any mixed strategy is a weighted average of these numbers. Therefore, the best he can hope for is a payoff of 1 unit.

Player II lists the worst that can happen to him if he plays pure strategies:

	Worst Possibility
Column 1	5
Column 2	5
Column 3	1

Again figuring conservatively, the best he can hope for is a payoff of one unit. Accordingly, Player I will play Row 2 always and Player II will play

Column 3 always and if either player deviates he will lose out. The crucial factor is that g_{23} is simultaneously the smallest element in its row and the largest element in its column. Such a point is called a *saddle* point. To check for a saddle point, simply circle the smallest element in each row, put a box around the largest element in each column and if any number is both circled and boxed, it is a saddle point. Thus

$$\mathbf{G} = \begin{pmatrix} \boxed{5} & \enclose{circle}{-2} & -1 \\ 3 & 2 & \boxed{\enclose{circle}{1}} \\ \enclose{circle}{-3} & \boxed{5} & -1 \end{pmatrix}$$

If g_{ij} is a saddle point, then Player I must play Row i, Player II must play Column j and $v = g_{ij}$.

Incidentally, it is possible to have two (or more) saddle points, in which case any mixture of the possible pure strategies is equally good. Hence, we should proceed as follows to analyze an arbitrary game:

1. Check for saddle points.
2. If no saddle, check for and strike out any dominated rows or dominating columns.
3. Use a random strategy on the reduced payoff matrix, if necessary.

For example,

$$\mathbf{G} = \begin{pmatrix} -1 & 3 & 1 \\ 1 & 2 & -1 \\ -2 & 4 & -2 \end{pmatrix}$$

Checking for saddle points, we have

$$\mathbf{G} = \begin{pmatrix} \enclose{circle}{-1} & 3 & \boxed{1} \\ \boxed{1} & 2 & \enclose{circle}{-1} \\ \enclose{circle}{-2} & \boxed{4} & \enclose{circle}{-2} \end{pmatrix}$$

We see that there is no saddle point but that we may strike Column 2, and having done so, we may strike Row 3:

$$G = \begin{pmatrix} -1 & \cancel{3} & 1 \\ 1 & \cancel{2} & -1 \\ \cancel{-2} & \cancel{4} & \cancel{-2} \end{pmatrix}$$

The remaining payoff matrix is

$$\mathbf{G}' = \begin{pmatrix} -1 & 1 \\ 1 & -1 \end{pmatrix}$$

We must now try a random strategy:

$$(p, 1-p)\begin{pmatrix} -1 & 1 \\ 1 & -1 \end{pmatrix} = (v, v)$$

Thus,

$$\begin{aligned} -p + 1 - p &= v \\ p - 1 + p &= v \end{aligned}$$

yielding

$$\begin{aligned} -2p + 1 &= 2p - 1 \\ 2 &= 4p \\ 1/2 &= p \end{aligned}$$

and hence $v = 0$. Similarly, Player II's strategy will yield $q = 1/2$. Thus, we finally arrive at a solution to $\mathbf{G}$:

$$\mathbf{p} = (1/2, 1/2, 0) \qquad \mathbf{q} = \begin{pmatrix} 1/2 \\ 0 \\ 1/2 \end{pmatrix}$$

and

$$v = 0$$

EXERCISES

1. Show that if for $\mathbf{G} = \begin{pmatrix} a & b \\ c & d \end{pmatrix}$ pure strategies can be arrived at by row and column dominance, then the resulting solution is also a saddle point, and conversely.

2. Solve the following games:

   (a) $\mathbf{G} = \begin{pmatrix} -1 & 1 \\ -3 & 0 \end{pmatrix}$

   (b) $\mathbf{G} = \begin{pmatrix} 1 & 0 & 3 \\ 2 & 1 & 1 \end{pmatrix}$

(c) $\mathbf{G} = \begin{pmatrix} 1 & -2 & -3 & 4 \\ 1 & -1 & 0 & 2 \\ 1 & -3 & 4 & -5 \end{pmatrix}$

3. Solve the following games:

(a) $\mathbf{G} = \begin{pmatrix} 2 & 8 & 3 & 4 \\ 4 & 3 & 4 & 4 \\ 6 & 6 & 5 & 5 \\ 7 & 8 & 6 & 5 \end{pmatrix}$

(b) $\mathbf{G} = \begin{pmatrix} 1 & 2 & 2 \\ 2 & 1 & 3 \\ 1 & -1 & -2 \end{pmatrix}$

4. Solve the following games:

(a) $\mathbf{G} = \begin{pmatrix} -3 & 1 & -2 \\ 5 & 4 & 2 \\ 0 & 3 & 3 \end{pmatrix}$

(b) $\mathbf{G} = \begin{pmatrix} -3 & 1 & -2 \\ 5 & 4 & 4 \\ 0 & 3 & 3 \end{pmatrix}$

5. Consider the card game in which the payoff to Player I is the product of the numbers if the colors match or to Player II if they do not match. Suppose Player I has a black 2, a black 3 and a red 3, and Player II has a black 1, a black 2 and a black 3. Find the payoff matrix and solve the game.

6. Mike suggests the following game to Harry: Each will pick a number 1, 2 or 3. If the numbers match Harry wins \$1. For nonmatching numbers, if either player plays a 3, he loses \$1. If either player plays a 2, Mike wins \$1. Should Harry agree to play? What is the best strategy for each player?

4.10 Larger Games

Let us consider some other possible payoff matrices. Let us assume that all the advantages of row and column domination have been realized and we are left with a $2 \times n$ game. It appears that Player II may select a mixed strategy which involves all n of his possible plays. However, it can be shown that his optimum strategy will turn out to use only two columns. So, for example,

$$\mathbf{G} = \begin{pmatrix} 1 & -1 & 3 \\ 3 & 4 & 0 \end{pmatrix}$$

Looking at only Column 1 and 2, we have

$$\mathbf{G}_1 = \begin{pmatrix} 1 & -1 \\ 3 & 4 \end{pmatrix}$$

For this game, there is a saddle point, so the best strategy is

$$\mathbf{q}_1 = \begin{pmatrix} 0 \\ 1 \end{pmatrix} \quad \text{and} \quad v_1 = 3$$

Looking at Columns 1 and 3, we have

$$\mathbf{G}_2 = \begin{pmatrix} 1 & 3 \\ 3 & 0 \end{pmatrix}$$

with optimal strategy $\mathbf{q}_2 = \begin{pmatrix} 3/5 \\ 2/5 \end{pmatrix}$ and $v_2 = 9/5$. Looking at Columns 2 and 3, we have

$$\mathbf{G}_3 = \begin{pmatrix} -1 & 3 \\ 4 & 0 \end{pmatrix}$$

which has optimal strategy, $\mathbf{q}_3 = \begin{pmatrix} 3/8 \\ 5/8 \end{pmatrix}$ and $v_3 = 3/2$. Clearly, the best for Player II is to play only Columns 2 and 3, so his best strategy is

$$\mathbf{q} = \begin{pmatrix} 0 \\ 3/8 \\ 5/8 \end{pmatrix}$$

Knowing this, Player I seeks his optimal strategy based only on $\mathbf{G}_3$ and gets $\mathbf{p} = (1/2, 1/2)$. Of course, $m \times 2$ games may be handled in an analogous manner.

Finally, let us consider a 3×3 matrix game without a saddle point. By an analysis similar to that for 2×2 games, the optimal strategies frequently may be found by solving

$$\mathbf{pG} = \mathbf{v} \quad \text{and} \quad \mathbf{Gq} = \mathbf{w}$$

where

$$\mathbf{p} = (p_1, p_2, p_3) \qquad p_1 + p_2 + p_2 = 1$$

$$\mathbf{q} = \begin{pmatrix} q_1 \\ q_2 \\ q_3 \end{pmatrix} \qquad q_1 + q_2 + q_3 = 1$$

$$\mathbf{v} = (v, v, v) \quad \text{and} \quad \mathbf{w} = \begin{pmatrix} w \\ w \\ w \end{pmatrix}$$

However, it must turn out that $w = v$. As an example, let us consider the following: Suppose each player has a black 1, a red 2 and a black 3. Each plays a card. If the colors match, the payoff to the first player is the sum of the cards. If not, the sum is paid to the second player. Here we have

$$\mathbf{G} = \begin{array}{c} \\ \\ \\ \end{array}\begin{pmatrix} 2 & -3 & 4 \\ -3 & 4 & -5 \\ 4 & -5 & 6 \end{pmatrix}\begin{array}{l} B1 \\ R2 \\ B3 \end{array}$$
$$\qquad\quad B1 \quad R2 \quad B3$$

We try

$$(p_1, p_2, p_3)\begin{pmatrix} 2 & -3 & 4 \\ -3 & 4 & -5 \\ 4 & -5 & 6 \end{pmatrix} = (v, v, v)$$

This yields

$$2p_1 - 3p_2 + 4p_3 = v \qquad \textbf{(1)}$$
$$-3p_1 + 4p_2 - 5p_3 = v \qquad \textbf{(2)}$$
$$4p_1 - 5p_2 + 6p_3 = v \qquad \textbf{(3)}$$
$$p_1 + p_2 + p_3 = 1 \qquad \textbf{(4)}$$

or

$$v - 2p_1 + 3p_2 - 4p_3 = 0 \qquad \textbf{(1)}$$
$$v + 3p_1 - 4p_2 + 5p_3 = 0 \qquad \textbf{(2)}$$
$$v - 4p_1 + 5p_2 - 6p_3 = 0 \qquad \textbf{(3)}$$
$$p_1 + p_2 + p_3 = 1 \qquad \textbf{(4)}$$

This may be solved by triangularization as follows:

$$\left(\begin{array}{cccc:c} 1 & -2 & 3 & -4 & 0 \\ 1 & 3 & -4 & 5 & 0 \\ 1 & -4 & 5 & -6 & 0 \\ 0 & 1 & 1 & 1 & 1 \end{array}\right) \sim \left(\begin{array}{cccc:c} 1 & -2 & 3 & -4 & 0 \\ 0 & 5 & -7 & 9 & 0 \\ 0 & -2 & 2 & -2 & 0 \\ 0 & 1 & 1 & 1 & 1 \end{array}\right)$$

$$\sim \left(\begin{array}{cccc:c} 1 & -2 & 3 & -4 & 0 \\ 0 & 1 & 1 & 1 & 1 \\ 0 & -2 & 2 & -2 & 0 \\ 0 & 5 & -7 & 9 & 0 \end{array}\right) \sim \left(\begin{array}{cccc:c} 1 & -2 & 3 & -4 & 0 \\ 0 & 1 & 1 & 1 & 1 \\ 0 & 0 & 4 & 0 & 2 \\ 0 & 0 & -12 & 4 & -5 \end{array}\right)$$

$$\sim \left(\begin{array}{cccc:c} 1 & -2 & 3 & -4 & 0 \\ 0 & 1 & 1 & 1 & 1 \\ 0 & 0 & 4 & 0 & 2 \\ 0 & 0 & 0 & 4 & 1 \end{array}\right)$$

Hence, $4p_3 = 1$, $p_3 = 1/4$; $4p_2 = 2$, $p_2 = 1/2$; $p_1 + p_2 + p_3 = 1$, $p_3 = 1/4$ and $v = 0$. Thus, $\mathbf{p} = (1/4, 1/2, 1/4)$ and $v = 0$. We could proceed similarly to find

$$\mathbf{q} = \begin{pmatrix} 1/4 \\ 1/2 \\ 1/4 \end{pmatrix}$$

but due to the symmetry of the game, it should be obvious that both players should use the same strategy.

EXERCISES

1. Solve the following games:

   (a) $\begin{pmatrix} 1 & -1 & 4 & 2 \\ 2 & 3 & 1 & 0 \end{pmatrix}$

   (b) $\begin{pmatrix} 1 & -2 & 4 \\ -1 & 2 & 0 \\ -1 & 1 & -1 \\ 3 & 0 & -3 \end{pmatrix}$

2. Player I has a black 2, a black 3 and a red 3. Player II has a red 1, a black 2 and a black 3. Each player plays a card. If the colors match, the payoff to Player I is the sum of the numbers; if not, the payoff is to Player II. Find the payoff matrix and show that the method in the text does not give a solution.

3. The two players have the same cards as in Exercise 2. However, the payoff to Player I is the sum of the numbers if the colors match and the positive difference if they do not. Find the payoff matrix and solve the game.

4. Joe and Sam agree to the following game. Each will pick a number 1, 2 or 3. If the numbers do not match, Sam wins \$1. If both pick 1, no one wins and if they match 2's or 3's, Joe wins \$3. Find the best strategies for both players.

5. Solve the following matrix games:

   (a) $\mathbf{G} = \begin{pmatrix} 7 & 5 \\ 8 & 4 \\ 7 & 4 \\ 1 & 2 \\ 2 & 9 \end{pmatrix}$

(b) $\mathbf{G} = \begin{pmatrix} 5 & 3 \\ 7 & 5 \\ 0 & 2 \\ 4 & 4 \end{pmatrix}$

(c) $\mathbf{G} = \begin{pmatrix} 0 & 1 & 3 & 6 \\ 9 & 4 & 0 & 6 \end{pmatrix}$

(d) $\mathbf{G} = \begin{pmatrix} 5 & 5 & 6 \\ 1 & 1 & 0 \\ 8 & 3 & 9 \end{pmatrix}$

(e) $\mathbf{G} = \begin{pmatrix} 9 & 1 & 6 \\ 7 & 6 & 8 \\ 7 & 9 & 8 \end{pmatrix}$

(f) $\mathbf{G} = \begin{pmatrix} 0 & 1 & -7 & 9 \\ 5 & 0 & 4 & 3 \\ 9 & 4 & 3 & 4 \end{pmatrix}$

(g) $\mathbf{G} = \begin{pmatrix} 3 & 6 & 1 & 5 \\ 9 & 0 & 8 & 0 \\ 7 & 5 & 2 & 6 \\ 7 & 7 & 9 & 5 \end{pmatrix}$

5

Conditional Probability and Matrices

5.1 Introduction

In many probability problems, one is asked to find the probability of some event B given that another event A has already occurred. We shall use the symbol $P(B \mid A)$, which is read "the probability of B given A," to stand for this measure. Sometimes the given is stated explicitly, but other times it must be deduced. For example, in a math course, seven girls passed, three girls failed, nine boys passed and four boys failed. The makeup of the class may be tabulated as

	P	F
B	9	4
G	7	3

Let us ask some simple questions. If a student is chosen at random, what is the probability that the student is a boy? The answer is easy, $P(B) = 13/23$. If a boy is chosen at random, what is the probability that he failed? Here we really have $P(F \mid B) = 4/13$, since we look only at that portion of the sample space which is composed of boys. Let us try the following: If a student is chosen at random, what is the probability that it is a boy who failed? Here we have $P(F \cap B) = 4/23$. Note the relationship

$$P(F \cap B) = P(B)P(F \mid B) \quad \text{or} \quad P(F \mid B) = \frac{P(F \cap B)}{P(B)}$$

If we think of probability as area in a Venn diagram, then we would want to say that the $P(F \mid B)$ is that fraction of the area of B which is also in F, leading us again to

$$P(F \mid B) = \frac{P(F \cap B)}{P(B)}$$

Let us take this formula as the definition of conditional probability.

Definition 1 $P(B \mid A) = P(B \cap A)/P(A)$ if $P(A) \neq 0$.

Interestingly, this formula is commonly used in the form $P(A \cap B) = P(A)P(B \mid A)$. For example, two cards are chosen at random from an ordinary deck of playing cards. What is the probability that both are hearts? Let A = the first card is a heart and B = the second card is a heart. We want $P(A \cap B)$. We write $P(A \cap B) = P(A)P(B \mid A)$. Now, $P(A) = 13/52$; $P(B \mid A) = 12/51$ since one card is used up and it is a heart. Hence $P(A \cap B) = 13/52 \cdot 12/51 = 1/17$.

Another common use of conditional probability is in computing the probability of complicated events which may be broken down into a collection of disjoint events. To be precise let $B_1, B_2, \ldots, B_k$ be a collection of k events having the following properties:

$$B_1 \cup B_2 \cup \ldots \cup B_k = S \tag{1}$$

$$B_i \cap B_j = \varnothing \quad \text{if } i \neq j \tag{2}$$

That is, the B's are disjoint events which together make up the whole sample space. Now, let A be any event; the subsets $A \cap B_1, A \cap B_2, \ldots, A \cap B_k$ partition A into disjoint events whose union is A. That is,

$$A = (A \cap B_1) \cup (A \cap B_2) \cup \ldots \cup (A \cap B_k) \tag{3}$$

$$(A \cap B_i) \cap (A \cap B_j) = \varnothing \quad \text{if } i \neq j \tag{4}$$

Thus, we can write

$$P(A) = P(A \cap B_1) + P(A \cap B_2) + \cdots + P(A \cap B_k)$$

and applying the conditional probability formula, we have

$$P(A) = P(A \mid B_1)P(B_1) + P(A \mid B_2)P(B_2) + \cdots + P(A \mid B_k)P(B_k)$$

Let us consider an example of the application of this formula. Suppose we have three urns with contents as shown:

Urn I	Urn II	Urn III
5 red balls	4 red balls	1 red ball
3 white balls	1 white ball	9 white balls

Let us refer to this as Urn Model A. An urn is chosen at random and a ball is selected from the urn at random. What is the probability that the ball is red? One might be tempted to say that there are 23 balls all together and 10 are red and hence the probability is 10/23. However, let us take a closer look. Let

$$\begin{aligned} R &= \text{A red ball is chosen.} \\ U_1 &= \text{Urn I is chosen.} \\ U_2 &= \text{Urn II is chosen.} \\ U_3 &= \text{Urn III is chosen.} \end{aligned}$$

Now,

$$P(U_1) = P(U_2) = P(U_3) = 1/3$$
$$P(R \mid U_1) = 5/8 \qquad P(R \mid U_2) = 4/5 \quad \text{and} \quad P(R \mid U_3) = 1/10$$

Our partitioning formula would say

$$\begin{aligned} P(R) &= P(R \mid U_1)P(U_1) + P(R \mid U_2)P(U_2) + P(R \mid U_3)P(U_3) \\ &= 5/8 \cdot 1/3 + 4/5 \cdot 1/3 + 1/10 \cdot 1/3 = 5/24 + 4/15 + 1/30 \\ &= 61/120 \end{aligned}$$

An interesting question is the following: With the same set-up as previously, a red ball is chosen. What is the probability that it came from Urn III? Here we want $P(U_3 \mid R)$. Our definition says

$$P(U_3 \mid R) = \frac{P(U_3 \cap R)}{P(R)}$$

$P(R)$ was just found. $P(U_3 \cap R)$ may be found by

$$P(U_3 \cap R) = P(R \cap U_3) = P(U_3)P(R \mid U_3) = 1/3 \cdot 1/10 = 1/30$$

Hence,

$$P(U_3 \mid R) = (1/30) \div (61/120)$$
$$= 1/30 \cdot 120/61 = 4/61$$

In other words, the information that a red ball was selected has dropped the probability that Urn III was selected from 1/3 to less than 1/15! This is intuitively reasonable due to the extreme unlikeliness of picking a red ball from Urn III.

Perhaps the simplest way to visualize problems of the preceding type is through the use of "tree diagrams." In a tree diagram, one starts at some point and draws branches indicating each of the possible outcomes of the first stage of the experiment. Thus figure 1 shows the start of a tree diagram for the urn problem. The numbers on the branches are the probabilities of the outcomes. Now, from each terminus, again draw branches for each possibility. Figure 2 shows the completed tree diagram. The numbers on the second set of branches give the probabilities of those branches given their starting points. Thus, $5/8 = P(R \mid U_1)$, and so on. Notice the following points:

1. The sum of the probabilities of all branches emanating from any point is one.
2. The probability of any sequence of outcomes from start to a given terminus is the product of the probabilities along the branch.

Naturally, this process can be extended to multistage experiments.

Most of the properties of conditional probability can be exhibited nicely by urn models. In addition, many problems can be translated into

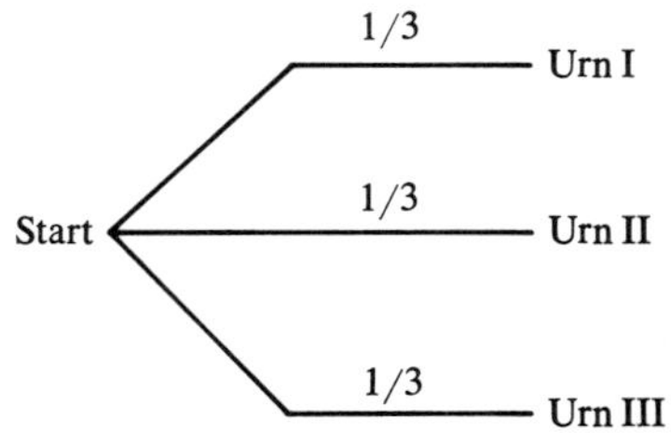

Figure 1

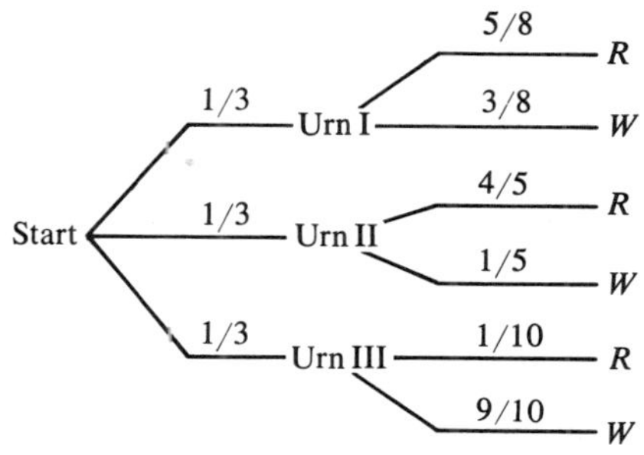

Figure 2

urn models. For example, ten men and five women volunteer for a special job. Two people are selected at random to do the job. What is the probability that the second person is a woman? This problem may be restated as Urn Model *B*. An urn contains ten black balls and five red balls. Two balls are drawn at random. What is the probability that the second ball is red? A tree provides the answer easily. Referring to figure 3, we find

$$P(R_2) = P(R_1 - R_2) + P(B_1 - R_2)$$
$$= 1/3 \cdot 4/14 + 2/3 \cdot 5/14 = 4/42 + 10/42 = 14/42 = 1/3$$

It was obvious that $P(R_1) = 1/3$ but should $P(R_2) = 1/3$? Think not of balls in an urn but of five red and ten black cards with identical backs, well-shuffled and laid out face down. If a card is turned over, what is the probability that it is red? The answer must be 1/3 whether it is the first card in line, or the second or the last. The same is true in drawing balls from an urn. Of course, if you know the color of the first one drawn, then the probabilities on the second one change. Conversely, knowledge about the second ball changes the probabilities on the first one.

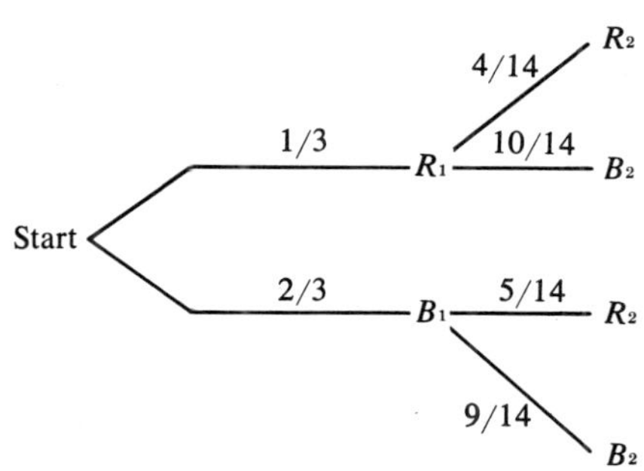

Figure 3

EXERCISES

1. In the Urn Model B example in the text, find the following:
   (a) the probability that the last ball picked is red;
   (b) the probability that the last ball picked is red given that the first one picked is red;
   (c) the probability that the first picked was red given that the last picked was red.

2. A scientist is running rats through a maze. If he ever succeeds, the probability of his succeeding again is 0.9. If he fails, the probability of succeeding is given by $k/10$ where k is the number of the trial, that is, $1/10$ on the first time, $2/10$ on the second, and so on. Draw a tree diagram for a rat with four tries at the maze. Find
   (a) the probability that the rat ever makes it;
   (b) the probability that the rat makes it on the last try;
   (c) the probability that the rat only makes it on his last try.

3. Find the probability that three cards drawn at random from a deck are all hearts. Find the probability of drawing a heart flush in five-card poker. Find the probability of drawing any flush.

4. A class consists of ten men and nine women. One-half of the men and one-third of the women smoke. What is the probability that a person chosen at random smokes? What is the probability that a smoker chosen at random is a woman?

5. There are two jars of nuts on a table. One contains ten peanuts and six cashews. The other contains five peanuts and twenty cashews. A jar is picked at random and a nut selected from the jar. What is the probability that it is a peanut? Another nut is chosen from the same jar; if the first nut was a peanut, what is the probability that the second nut is a peanut? If two peanuts were selected, what is the probability that they were selected from the second jar?

6. A poker player has two aces, two kings and a jack. He discards the jack and picks a card. What is the probability that he ends up with a full house?*

7. A jar contains four nickels and two dimes. Four coins are chosen successively at random. Find

---

* A full house is a five-card poker hand consisting of three cards of one denomination and two cards of another.

(a) the probability that the total drawn is 30¢;
(b) the probability that the first two coins drawn total more money than the last two;
(c) the probability that the coins alternate in denomination.

8. A box contains ten tags numbered 1, 2, 3, . . . , 10. Arthur draws a tag numbered 7 and keeps it. Now Sue picks a tag. What is the probability that Sue's number is higher than Arthur's?

9. The probability that I am late is 1/3. The probability of getting caught in the rain is 1/4. The probability of being late if I am caught in the rain is 1/2. Find
(a) the probability of being late and caught in the rain;
(b) the probability of being late and not caught in the rain;
(c) the probability of being late or caught in the rain;
(d) the probability that I was caught in the rain given that I am late.

10. $P(A \mid B) = 1/2$, $P(A) = 1/6$ and $P(B \mid A) = 1/12$. Find $P(B)$.

11. $P(A \cup B) = 2/3$, $P(A) = 1/2$, $P(B) = 7/12$. Find $P(B \mid A)$ and $P(A \mid B)$.

5.2 Bayes' Theorem

Normally when one constructs a tree, one starts at the first stage and builds towards the end, using conditional probabilities. Sometimes, however, it is required to find the first probabilities conditioned upon knowledge of the latter stages. The formula by which one works backward in this way is called *Bayes' theorem.* We have actually used this theorem in the last section without naming it. The theorem supposes that an event A is attained on the second (or later) stage of an experiment and can be reached via k events $B_1, B_2, \ldots, B_k$ on the first stage. It is assumed that the B's form a partition of S, so that

$$P(A) = P(A \mid B_1)P(B_1) + P(A \mid B_2)P(B_2) + \cdots + P(A \mid B_k)P(B_k)$$

The values $P(B_1), P(B_2), \ldots, P(B_k)$ are assumed to be known. Now, if A occurs, what is $P(B_j \mid A)$ where $j = 1, 2, \ldots, k$? The answer is

$$P(B_j \mid A) = \frac{P(A \cap B_j)}{P(A)} = \frac{P(A \mid B_j)P(B_j)}{P(A)} \qquad j = 1, 2, \ldots, k$$

where $P(A)$ is given above. The numerator of the fraction is simply the probability of the branch $B_j - A$ on the tree for the experiment.

Many statisticians like to reason in this Bayesian fashion. That is, they assume values for $P(B_1)$, $P(B_2)$, . . . , $P(B_k)$ which they believe are good guesses based on past experience. These are called *a priori* or *prior* probabilities. After the experiment is performed, the computed $P(B_1 \mid A)$, $P(B_2 \mid A)$, . . . , $P(B_k \mid A)$ are now called the *a posteriori* or *posterior* probabilities. If another experiment is run, they will use these new probabilities as the prior probabilities. For example, let us try Urn Model C. Urn 1 contains one red and nine black balls. Urn II contains seven red and three black balls. An urn is chosen at random. A ball is chosen from the urn and it is red. What is the probability that Urn I was chosen? From figure 4,

$$P(U_1 \mid R) = \frac{P(U_1 \cap R)}{P(R)} = \frac{(0.5)(0.1)}{(0.5)(0.1) + (0.5)(0.7)} = \frac{0.05}{0.40} = \frac{1}{8}$$

Now, it must be that $P(U_2 \mid R) = 7/8$.

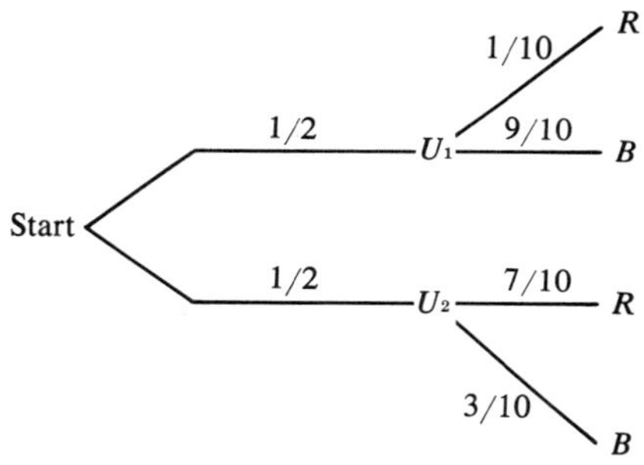

Figure 4

Suppose that we now replace the ball, shake up the urn and draw again. Suppose the ball is again red. What is the probability that Urn I was chosen? In figure 5, we have the same diagram except that the prior probabilities of the urns have been adjusted to take into account the previous information. Now,

$$P(U_1 \mid R) = \frac{P(U_1 \cap R)}{P(R)} = \frac{(1/8)(1/10)}{(1/8)(1/10) + (7/8)(7/10)} = \frac{1}{50}$$

Therefore, $P(U_2 \mid R) = 49/50$.

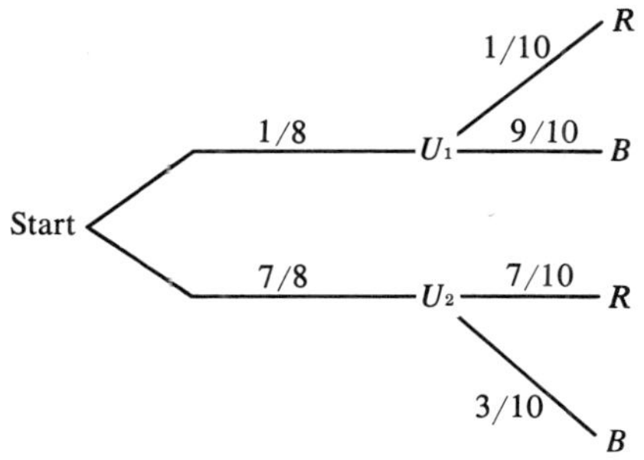

Figure 5

Interestingly, we could have considered this problem as two draws from the urn starting with the original prior, as shown in figure 6. From this diagram we see

$$P(U_1 \mid R - R) = \frac{P(U_1 - R - R)}{P(R - R)} = \frac{(0.5)(0.1)(0.1)}{(0.5)(0.1)(0.1) + (0.5)(0.7)(0.7)}$$

$$= \frac{0.005}{0.005 + 0.245} = \frac{1}{50}$$

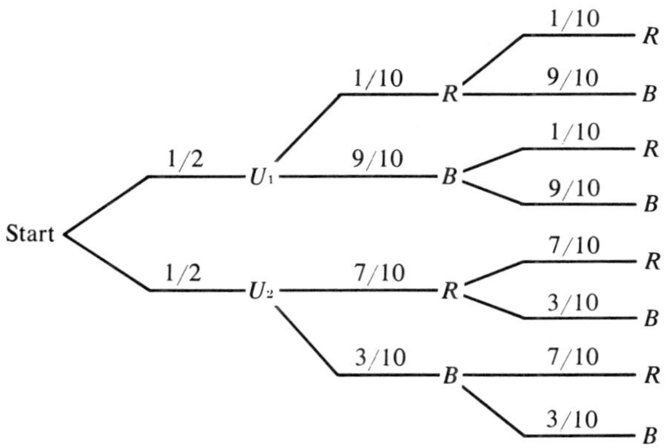

Figure 6

EXERCISES

1. Urn I contains six white and three green balls. Urn II contains two white balls and seven green balls. An urn is chosen at random and a ball is chosen from it at random. Given that the ball was white, find the posterior probabilities of selecting the two urns.

2. Box 1 contains five slips with the number 1 and three with the number 2. Box 2 contains six slips with the number 1 and six with the number 2. A box is chosen at random and a slip selected from the box. The number on the slip tells which box to pick the next slip from. If the second slip is numbered 2, find the probability that the first slip was numbered 1. Find the probability that the first box chosen was number 1. (Draw a tree diagram.)

3. In a good teacher's math class, 10% of the students fail. In a bad teacher's math class, 20% of the students fail. Only 5% of the teachers are bad teachers. If a student selected at random passed math, what is the probability that he had a good teacher?

4. The probability that a willow will wither if it is watered is 1/3. The probability that a willow will wither if it is not watered is 1/2. The probability that Walter will water the willow is 3/4. The willow withered. What is the probability that Walter watered it?

5. Of all potatoes, 1/4 are small, 1/4 are large and 1/2 are medium-sized. Of small potatoes, 10% are rotten; of medium potatoes, 15% are rotten; and of large potatoes, 40% are rotten. If a potato chosen at random is rotten, find the posterior probabilities of its possible sizes.

6. An observation is made of a random variable which may be distributed according to $p_1(x)$, $p_2(x)$ or $p_3(x)$, given in Section 4.7. A priori, all three distributions are equally likely. The value observed is $x = 5$. Find the posterior probabilities of the three distributions.

7. Continuation of 6. A second observation is made and it is $x = 4$. What are the posterior probabilities of the three distributions now?

8. (a) Assuming that the first observation was $x = 3$, repeat Exercise 6.
   (b) Another observation is made, again $x = 3$. What are now the posterior probabilities?

5.3 Independence

Let us reconsider Urn Model *C*. Recall that Urn I contains one red ball and nine black balls, Urn II contains seven red and three black balls. A ball chosen from an urn selected at random is red. The posterior probabilities of U_1 and U_2 are 1/8 and 7/8, respectively. Suppose a second ball is drawn without replacement and it is red. According to our analysis of Urn Model *B*, the probability of R_2 and R_1 is the same. Hence, we have a tree as shown in figure 7a.

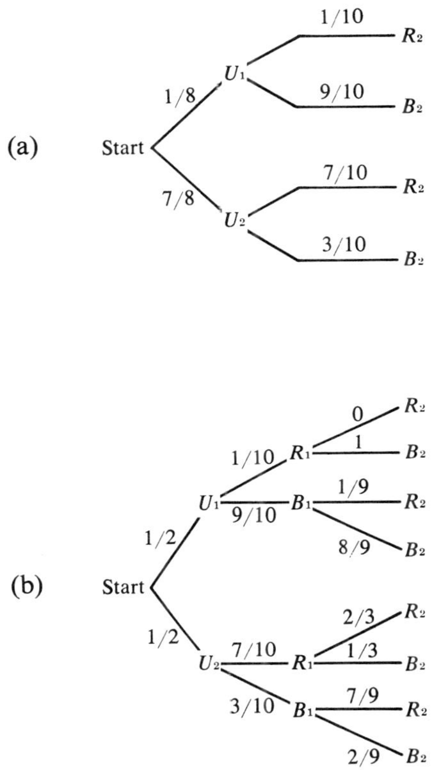

$$P(U_1 \mid R_2) = \frac{P(U_1 \cap R_2)}{P(R_2)} = \frac{0}{2/5} = 0$$

Figure 7

From the diagram, $P(R_2) = 1/80 + 49/80 = 5/8$ and

$$P(U_1 \mid R_2) = \frac{P(U_1 \cap R_2)}{P(R_2)} = \frac{1/80}{5/8} = \frac{1}{50}$$

However, clearly since both picks were red, the posterior probability of U_1 must be zero since U_1 had only one red ball and the sampling was done without replacement. The difficulty is caused by the fact that $P(R_2 \mid R_1) \neq P(R_2)$. The correct tree is shown in figure 7b. When the sampling was with replacement, $P(R_2 \mid R_1) = P(R_2)$, that is, knowing that R_1 occurred did not change the probability of R_2 occurring. We express this idea by saying that R_2 is independent of R_1. So if A is independent of B, then $P(A \mid B) = P(A)$. In general, $P(A \mid B) = P(A \cap B)/P(B)$. If we have independence, then $P(A) = P(A \cap B)/P(B)$, or

$$P(A \cap B) = P(A)P(B)$$

In this form, the expression is symmetric in A and B. That is, if A is independent of B, then B is independent of A. Therefore, we shall take as a definition the following.

Definition 2 A and B are called independent if $P(A \cap B) = P(A)P(B)$.

Of course, if in a description of an experiment different events appear to be totally unrelated, we may use this definition to compute probabilities. For example, two fair coins are flipped. What is the probability that both come up heads? Let H_1 be the event "the first coin is heads," and H_2 be the event "the second coin is heads." Certainly, the outcome on one coin should not affect the other. Hence, $P(H_1 \cap H_2) = P(H_1)P(H_2) = (1/2)(1/2) = 1/4$.

It is interesting that given three events A, B and C, it is possible that C is independent of A, and C is independent of B but C is *not* independent of $A \cap B$. For example, a box contains four cards. One card is blue, one card is green, one card is red and the last card is paisley–red, green and blue. A card is drawn at random. Let B = a card with blue on it; G = a card with green on it; and R = a card with red on it. Now, $P(R) = P(G) = P(B) = 1/2$. R is independent of both G and B, since $P(R \cap G) = 1/4 = P(R)P(G)$ and $P(R \cap B) = 1/4 = P(R)P(B)$. But $P(R \mid B \cap G) = 1$ since given B and G, the only possible draw is the paisley card and so R must also have occurred. In addition, B and G are also independent. Therefore, the events B, G and R would be called *pairwise independent* but you would surely not want to say they form an independent threesome. Accordingly, we have the following definition.

Definition 3 A collection of events $A_1, A_2, \ldots, A_k$ are called *independent* if for every possible subcollection, the probability of the intersection is the product of the probabilities. For example, A_1, A_2, A_3 and A_4 are called independent only if they are pairwise independent and, in addition,

$$P(A_1 \cap A_2 \cap A_3) = P(A_1)P(A_2)P(A_3)$$
$$P(A_1 \cap A_2 \cap A_4) = P(A_1)P(A_2)P(A_4)$$
$$P(A_1 \cap A_3 \cap A_4) = P(A_1)P(A_3)P(A_4)$$
$$P(A_2 \cap A_3 \cap A_4) = P(A_2)P(A_3)P(A_4)$$
$$P(A_1 \cap A_2 \cap A_3 \cap A_4) = P(A_1)P(A_2)P(A_3)P(A_4)$$

Of course, tree diagrams for sequences of independent events are particularly simple to construct. Consider the following case: It has been found that 80% of all children exposed to chicken pox will come down with the disease. If three children are exposed to the infection, what is the probability that exactly two will contract it? Let C_1 be the event "the first child contracts the disease," C_1' be the event "the first child does not get sick" and so on for C_2, C_2', C_3 and C_3'. Figure 8 shows the tree diagram for this situation and

$$\begin{aligned} P(\text{Exactly } 2) &= P(C_1 - C_2' - C_3) + P(C_1 - C_2 - C_3') + P(C_1' - C_2 - C_3) \\ &= (0.8)(0.2)(0.8) + (0.8)(0.8)(0.2) + (0.2)(0.8)(0.8) \\ &= (3)(0.2)(0.8)(0.8) = 0.384 \end{aligned}$$

This is an example of a so-called binomial distribution which results from independent repetition of trials in which the outcomes can be classified as either "success" or "failure." We shall discuss the general problem in Section 7.3.

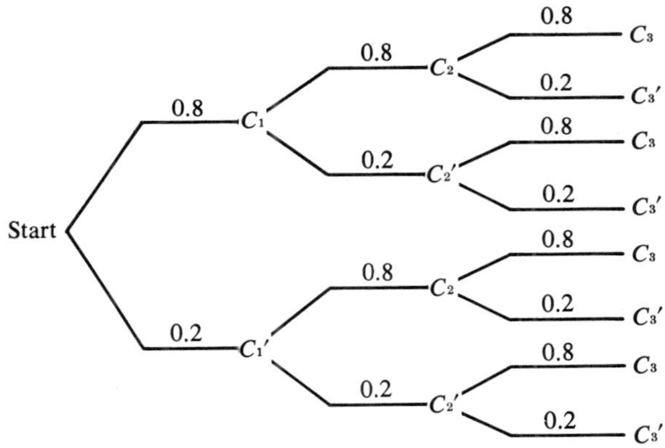

Figure 8

EXERCISES

1. A and B are independent events, $P(A) = 0.7$, $P(B) = 0.8$. What is $P(A \cap B)$?
2. A and B are independent events, $P(A) = 0.6$ and $P(A \cap B) = 0.2$. What is $P(A \cup B)$?
3. A and B are independent events, $P(A \cup B) = 0.8$ and $P(A) = 0.6$. What is $P(B)$?

4. Show that if A and B are independent, then so are A and B', A' and B, and A' and B'.

5. A and B are independent events, $P(A \cup B) = 0.76$ and $P(A \cup B') = 0.94$. Find $P(A)$ and $P(B)$.

6. $P(A \cup B) = 0.8$, $P(A) = 0.7$ and $P(B) = 0.6$. Are A, B independent? Why or why not?

7. An urn contains six white and three red balls. Three balls are chosen successively at random. Draw tree diagrams for this situation assuming first that the drawing is without replacement and second with replacement. Compare the probabilities of drawing three red balls under the two assumptions.

8. The probability that a man is shorter than 64 inches is 1/3. Find the probability that of four men chosen at random, exactly two are shorter than 64 inches.

9. Use the information in Exercise 8 to find the probability that at least one man out of four is shorter than 64 inches.

10. A, B and C are independent events. $P(A \cap B \cap C) = 0.2$, $P(A) = 0.8$ and $P(A \cap B) = 0.6$. Find $P(B)$, $P(C)$, $P(B \cap C)$, $P(A \cap C)$ and $P(A \cup B \cup C)$.

11. Mr. Able, Mr. Black and Mr. Clark check their hats. The hat check operator gives them back randomly. What is the probability that at least one man gets his own hat back? Are the events "Mr. Able gets his hat back" and "Mr. Black gets his hat back" independent? Explain.

5.4 Stochastic Processes

Let us consider the following learning model. A scientist is training a rat to run through a maze. The probability of the rat being successful is $(k + 1)/(n + 2)$ where n is the number of times the rat has tried the maze and k is the number of times the rat has succeeded. Figure 9 shows a tree diagram for three trials.

Let X_1 = number of successes after one trial. The range of X_1 is $\{0, 1\}$. The probability function for X_1, $p_1(x)$ is given by

x	$p_1(x)$
0	1/2
1	1/2

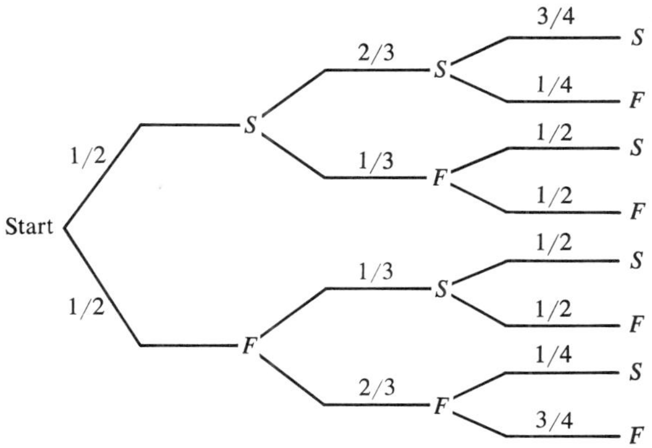

Figure 9

Let X_2 = number of successes after two trials. The range of X_2 is $\{0, 1, 2\}$. The probability function for X_2, $p_2(x)$ is given by

x	$p_2(x)$
0	1/3
1	1/3
2	1/3

Do you think that it is generally true that X_r would have range $\{0, 1, 2, \ldots, r\}$ and that $p_r(j) = 1/(r + 1)$ for $j = 0, 1, 2, \ldots, r$? Could you prove it? It happens to be true. More importantly, we have here an example of a sequence of random variables, $X_1, X_2, \ldots, X_r, \ldots$. Each random variable has a probability distribution, and the distribution depends upon the number of the trial. If we think of the number of trials as measuring time passing, then the different probability distributions may be thought of as a single distribution which varies as a function of time. In this case, the probability of having j successes in r tries is $1/(r + 1)$ and the distribution as a function of time is very simple. Such a mathematical structure is called a *stochastic process.*

Let us take another urn model as an example. Urn I contains eight tags numbered 1 and eight tags numbered 2. Urn II contains four tags numbered 1 and twelve tags numbered 2. To start the process, a tag is selected at random from Urn I. The number on the tag is noted. For the second step, a tag is chosen from the urn whose number was on the first tag and so on. A tag once chosen is not replaced in the urn. The process ends when there are no tags left to be selected in the appropriate urn. At each step the possible numbers that may be drawn are 1 and 2. We shall refer to these as the "states" of the process. By means of a tree diagram, the reader should ascertain that on the first pick, $P(\text{State } 1) = 1/2$ and $P(\text{State } 2) = 1/2$. On the second pick, $P(\text{State } 1) = 43/120$ and $P(\text{State } 2) = 77/120$, etc.

Let us take a second learning model as an example of a stochastic process. Again, suppose a scientist is training a rat to run a maze. The probability that the rat does not succeed on a given trial is $1/2^{k+1}$ where k is the number of previous successes. A tree diagram would look like figure 10 for the first few trials. As in the first example, we let the "state" of the process at the rth step be X_r = number of successes in r attempts.

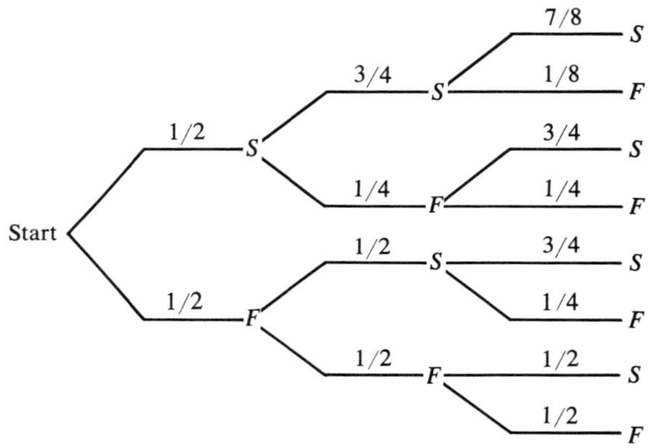

Figure 10

Now, the difference between this last model and the preceding two is that if one is told the state of the process at the rth step, then one can say exactly the possible states on the $(r + 1)$st step and the probabilities of entering into each of these states independent of r and of how the state X_r was reached. That is, if you are told that the rat has had four previous successes, the probability of having four successes after the next trial is $1/2^5 = 1/32$ and the probability of having five successes is $31/32$; and nothing else is possible regardless of how many steps have passed. This type of process is called a *Markov process* or *Markov chain*. In the first learning model, it would be necessary to know not only the number of successes, but also how long the process has been going on. In the urn model, if you were told that the process was in State 2, you would have no idea what the probabilities of going into State 1 or State 2 are without having a good deal of knowledge about the history of the process. In other words, a Markov process may be thought of as a filmstrip which is running on. You observe one frame of the strip without seeing any others and yet you are able to state with absolute assurance the likelihoods of everything that may be on the next frame. Strangely enough, this "memoryless" process can be used as a model for all kinds of practical problems.

The only thing annoying about our second learning model is that there is no limit to the number of possible states (i.e., number of possible successes) if the process goes on long enough. Thus, we would have to say that the number of possible states is infinite. Besides taking us out of the realm of finite mathematics, this causes difficulties which would be nice to avoid. Hence, in the next section, we shall discuss Markov processes in which there are only a fixed number of possible states. These so-called finite Markov chains turn out to have vast applications, of which only a few will be indicated.

EXERCISES

1. As a miniature version of the urn model with numbered tags, suppose the first urn contains one tag numbered 1 and one tag numbered 2, and the second urn contains one number 1 and two number 2's.
   (a) Assuming that the first pick is from Urn I, construct a tree diagram showing all possibilities.
   (b) What is the probability that the final pick is a 1?
   (c) What is the probability that the third pick is a 1?
   (d) What is the expected number of picks? What is the probability that the second pick was a 1 given that the third pick was a 2?
2. Assuming that the first pick is from Urn II, repeat Exercise 1.
3. Assuming that the first urn to be picked from is selected at random, repeat Exercise 1.
4. Consider a learning model in which the probability of a failure is $1/(k + 2)$ where k is the number of successes previously. Draw a tree diagram for the first three trials. What are the probabilities of none, one, two and three successes? What is the expected number of successes? What is the probability that there was a success on the first trial given that there was a success on the third?
5. Repeat Exercise 4 for both learning models given in the text.
6. For the first learning model in the text, suppose the process ends with the third success. What is the probability that the process ends in four steps? What is the probability that the process ends in less than five steps?

7. Repeat Exercise 6 for the second learning model in the text.
8. Repeat Exercise 6 for the model in Exercise 4.
9. As an example of a process with a short memory, consider the following: For a hockey team a win represents two points; a tie, one point; and a loss, no points. If the team has just won a game, it has probability 1/2 of winning, 1/4 of tying and 1/4 of losing. If it has just lost, the probabilities are 1/4 of winning, 1/4 of tying and 1/2 of losing. If it has just tied, the probabilities are 1/3 for each. Draw a tree for a three-game series following a tie.
   (a) Given that the first game was won, what is the probability of winning one out of the next two?
   (b) Given that the team won one out of the last two, what is the probability that it won the first game?
   (c) If the team won the first game, what is the expected number of wins in the next two?

5.5 Finite Markov Chains

At the end of the last section we defined a finite Markov chain. Let us review that definition. To begin, a finite Markov chain is a stochastic process having a finite number of states, say $s_1, s_2, \ldots, s_k$. At each step in the process there are probabilities of going from the existing state into other states. These we shall call *transition* probabilities. In addition, the process has no memory, that is, given that the process is in state s_i, the probability of its going into state s_j, on the next step, is p_{ij}; and p_{ij} is independent of the history of the process. Thus, all the information about the process is given by the numbers p_{ij}; $i = 1, 2, \ldots, k, j = 1, 2, \ldots, k$. As you can guess from the notation, we shall display the p_{ij}'s as a matrix **P**, called the *transition* matrix. A typical 4×4 transition matrix might be

$$\mathbf{P} = \begin{pmatrix} 1/2 & 1/4 & 1/8 & 1/8 \\ 1/3 & 1/3 & 0 & 1/3 \\ 1/4 & 1/2 & 0 & 1/4 \\ 1 & 0 & 0 & 0 \end{pmatrix}$$

Let us try to ascertain the meanings of some of these numbers. Take the first row. These are the numbers $p_{11}, p_{12}, p_{13}, p_{14}$. They represent the probabilities of going from state s_1 into s_1, s_2, s_3 and s_4 respectively. Notice that these fractions add up to one since all the possible transitions from s_1 are covered. Since the other rows cover the possible transitions from s_2, s_3 and

s_4, the sums of these rows must also be one. This is, of course, generally true for the rows of any transition matrix. We see that $p_{23} = 0$, which means that it is impossible to get from state s_2 to state s_3. Also, $p_{41} = 1$ and $p_{42} = p_{43} = p_{44} = 0$. Thus, whenever this process reaches state s_4, it must go into s_1 on the next step.

Let us note that any transition matrix must have the same number of rows as it has columns. Such a matrix is called a *square* matrix. We shall take the following definition.

Definition 4 Every square matrix with nonnegative entries and such that the sum of the numbers in each row is one shall be called a *stochastic* matrix and it represents the transition matrix for some Markov process.

Consider the given matrix **P**. Suppose that at the kth step we are told that $P(s_1) = 1/2$, $P(s_2) = 1/4$, $P(s_3) = 1/6$ and $P(s_4) = 1/12$. What are the probabilities of these states on the $(k + 1)$st step? Let us draw a tree (figure 11). We now can read

$$P(s_1) = (1/2)(1/2) + (1/4)(1/3) + (1/6)(1/4) + (1/12)(1)$$

If we had written the original probabilities of s_1, s_2, s_3 and s_4 as a vector

$$\mathbf{v} = (1/2, 1/4, 1/6, 1/12)$$

we see that the probability of being in state s_1 on the next step is the product of **v** with the first column of **P**. Similarly, the product of **v** with

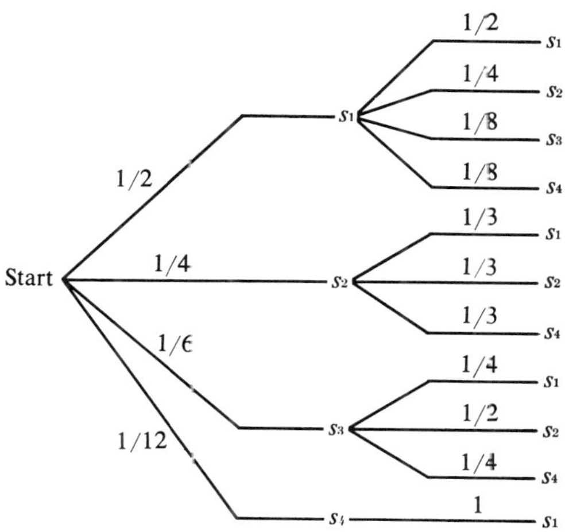

Figure 11

the second column of **P** is the new probability of s_2, and so on. Hence the product **vP** gives a row vector, the entries of which are the probabilities of being in the various states on the next step. Let us denote by $\mathbf{v}_0$ the initial vector of probabilities. Now, $\mathbf{v}_1 = \mathbf{v}_0\mathbf{P}$ is the vector of probabilities on the next step. Similarly, $\mathbf{v}_2 = \mathbf{v}_1\mathbf{P} = (\mathbf{v}_0\mathbf{P})\mathbf{P}$ is the vector of probabilities after two steps and so forth.

In general, let **P** be the transition matrix for a k state Markov process. That is,

$$\mathbf{P} = \begin{pmatrix} p_{11} & p_{12} & \cdots & p_{1k} \\ p_{21} & p_{22} & \cdots & p_{2k} \\ \vdots & & & \vdots \\ p_{k1} & p_{k2} & \cdots & p_{kk} \end{pmatrix}$$

Let $\mathbf{v}_0 = (a_1, a_2, \ldots, a_k)$ be the vector of probabilities of the states at some given time. Let $\mathbf{v}_1 = (b_1, b_2, \ldots, b_k)$ be the vector of probabilities on the next step. That is, $b_j = P(s_j \text{ after one step})$. To calculate b_j we write

$$\begin{aligned} b_j &= P(s_j \mid s_1 \text{ now})P(s_1 \text{ now}) + P(s_j \mid s_2 \text{ now})P(s_2 \text{ now}) \\ &\quad + \cdots + P(s_j \mid s_k \text{ now})P(s_k \text{ now}) \\ &= P(\text{going from } s_1 \text{ to } s_j)a_1 + P(\text{going from } s_2 \text{ to } s_j)a_2 \\ &\quad + \cdots + P(\text{going from } s_k \text{ to } s_j)a_k \\ &= a_1p_{1j} + a_2p_{2j} + \cdots + a_kp_{kj} \end{aligned}$$

Thus, b_j is the product of $\mathbf{v}_0$ and the jth column of P. But this is true for $b_1, b_2, \ldots, b_k$. Therefore

$$\mathbf{v}_0\mathbf{P} = \mathbf{v}_1$$

Take another example. Suppose we classify people by whether they own luxury cars (State 1), intermediate priced cars (State 2), and economy cars (State 3). Now most people try to move upwards, so each year there is a possible transition from state to state. We let

$$\mathbf{P} = \begin{pmatrix} .7 & .3 & 0 \\ .5 & .3 & .2 \\ 0 & .5 & .5 \end{pmatrix}$$

give the probabilities of the transitions in any one year. An independent survey shows that this year 60% of the population own economy cars, 30% own intermediate cars and 10% own luxury cars. If the transition probabilities remain stable, what will the percentages look like in two years? Now, $\mathbf{v}_0 = (0.1, 0.3, 0.6)$. (Note "luxury" is State 1.) After one year we have,

$$\mathbf{v}_1 = \mathbf{v}_0\mathbf{P} = (.1, .3, .6)\begin{pmatrix} .7 & .3 & 0 \\ .5 & .3 & .2 \\ 0 & .5 & .5 \end{pmatrix}$$
$$= (.22, .42, .36)$$

After two years we have

$$\mathbf{v}_2 = \mathbf{v}_1\mathbf{P} = (.22, .42, .36)\begin{pmatrix} .7 & .3 & 0 \\ .5 & .3 & .2 \\ 0 & .5 & .5 \end{pmatrix}$$
$$= (.364, .372, .264)$$

Thus, we see a marked trend away from economy and toward luxury! This is only a hypothetical example.

It would be nice to know if there is some way to get from $\mathbf{v}_0$ to $\mathbf{v}_2$ directly without passing through $\mathbf{v}_1$. That is, we would like to produce a matrix $\mathbf{P}^{(2)}$ which gives the two-step transition probabilities. Mathematically, $\mathbf{v}_2 = \mathbf{v}_0\mathbf{P}^{(2)}$ would give the same result as $\mathbf{v}_2 = \mathbf{v}_1\mathbf{P} = (\mathbf{v}_0\mathbf{P})\mathbf{P}$. In other words, $\mathbf{v}_0\mathbf{P}^{(2)} = (\mathbf{v}_0\mathbf{P})\mathbf{P}$. Of course, the ideal solution would be to define $\mathbf{P}$ times $\mathbf{P}$ in such a way that $\mathbf{PP} = \mathbf{P}^{(2)}$. This would give us as a fringe benefit that

$$(\mathbf{v}_0\mathbf{P})\mathbf{P} = \mathbf{v}_0(\mathbf{PP})$$

that is, the associative law. More generally, let

$$\mathbf{P}^{(n)} = \begin{pmatrix} a_{11} & a_{12} & \cdots & a_{1k} \\ a_{21} & a_{22} & \cdots & a_{2k} \\ \vdots & & & \vdots \\ a_{k1} & a_{k2} & \cdots & a_{kk} \end{pmatrix}$$

be the n-step transition matrix and let

$$\mathbf{P}^{(m)} = \begin{pmatrix} b_{11} & b_{12} & \cdots & b_{1k} \\ b_{21} & b_{22} & \cdots & b_{2k} \\ \vdots & & & \vdots \\ b_{k1} & b_{k2} & \cdots & b_{kk} \end{pmatrix}$$

be the m-step transition matrix for the process. Let us try to find $\mathbf{P}^{(n+m)}$, the $(n + m)$-step transition matrix. Let the entry in the ith row and jth column of $\mathbf{P}^{(n+m)}$ be c_{ij}. Now, c_{ij} is the probability that if we are now in state s_i we will be in state s_j in $n + m$ steps. The tree diagram of figure 12 shows the possible ways this can occur.

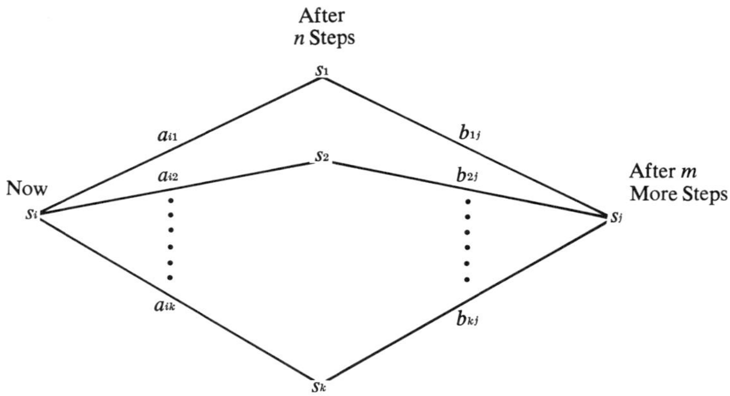

Figure 12

The probability of going from s_i to s_j is just the sum of the probabilities of each branch,

$$c_{ij} = a_{i1}b_{1j} + a_{i2}b_{2j} + \cdots + a_{ik}b_{kj}$$

Thus, c_{ij} is the product of the ith row of $\mathbf{P}^{(n)}$ and the jth column of $\mathbf{P}^{(m)}$. Let us define this to be the product of two matrices. A more precise definition follows.

Definition 5 Let **A** and **B** be two matrices. Then **AB** is a matrix **C** defined as follows: The entry in row i and column j of **C** is the product of the ith row of **A** with the jth column of **B**. If it is not possible to form the product of the ith row of **A** and the jth column of **B**, then the product is undefined.

Notice that the only time that one would not be able to form the product of a row and column vector is when they do not have the same number of components. For the time being, we are only dealing with $k \times k$ transition matrices and so each row and each column has k components and all the products can be formed. In particular, $\mathbf{P}^{(2)} = \mathbf{PP}$, $\mathbf{P}^{(3)} = \mathbf{P}^{(2)}\mathbf{P} = \mathbf{PP}^{(2)}$, $\mathbf{P}^{(4)} = \mathbf{PP}^{(3)} = \mathbf{P}^{(2)}\mathbf{P}^{(2)} = \mathbf{P}^{(3)}\mathbf{P}^{(1)}$ and so forth. Hence, we are justified in writing $\mathbf{P}^n$ for $\mathbf{P}^{(n)}$. That is, the n-step transition matrix is the nth power of **P** assuming that we perform the multiplications according to the rules set forth in the definition. Going back to our last example, we see that

$$\mathbf{P} = \begin{pmatrix} .7 & .3 & 0 \\ .5 & .3 & .2 \\ 0 & .5 & .5 \end{pmatrix}$$

$$\mathbf{P}^2 = \mathbf{PP} = \begin{pmatrix} .7 & .3 & 0 \\ .5 & .3 & .2 \\ 0 & .5 & .5 \end{pmatrix}\begin{pmatrix} .7 & .3 & 0 \\ .5 & .3 & .2 \\ 0 & .5 & .5 \end{pmatrix}$$

The entry in Row 1, Column 1 of $\mathbf{P}^2$ is the product

$$(.7 \quad .3 \quad 0)\begin{pmatrix} .7 \\ .5 \\ 0 \end{pmatrix} = .64$$

The entry in Row 1, Column 2 of $\mathbf{P}^2$ is the product

$$(.7 \quad .3 \quad 0)\begin{pmatrix} .3 \\ .3 \\ .5 \end{pmatrix} = .30$$

Continuing in like manner, we have

$$\mathbf{P}^2 = \begin{pmatrix} .64 & .30 & .06 \\ .50 & .34 & .16 \\ .25 & .40 & .35 \end{pmatrix}$$

Also,

$$\mathbf{P}^3 = \mathbf{P}^2\mathbf{P} = \begin{pmatrix} .64 & .30 & .06 \\ .50 & .34 & .16 \\ .25 & .40 & .35 \end{pmatrix}\begin{pmatrix} .7 & .3 & 0 \\ .5 & .3 & .2 \\ 0 & .5 & .5 \end{pmatrix}$$

$$= \begin{pmatrix} .598 & .312 & .090 \\ .520 & .332 & .148 \\ .375 & .370 & .255 \end{pmatrix}$$

Checking this out with all the decimals is annoying, but since probabilities must be numbers between zero and one, the alternative is fractions, which are even less appetizing. Nevertheless, the reader should check that

1. each matrix shown is a transition matrix as predicted;
2. $\mathbf{P}^2\mathbf{P} = \mathbf{PP}^2$, that is, $\mathbf{P}^3$ is well defined;
3. $\mathbf{v}_0\mathbf{P}^2 = (\mathbf{v}_0\mathbf{P})\mathbf{P}$ and $\mathbf{v}_0\mathbf{P}^3 = (\mathbf{v}_0\mathbf{P}^2)\mathbf{P}$, that is, the associative law holds.

EXERCISES

1. Determine which of the following are transition matrices. Find $\mathbf{P}^2$ for each.

(a) $\mathbf{P} = \begin{pmatrix} 0 & 1/2 \\ 1/3 & 2/3 \end{pmatrix}$

(b) $\mathbf{P} = \begin{pmatrix} 0 & 1 \\ 1 & 0 \end{pmatrix}$

(c) $\mathbf{P} = \begin{pmatrix} .5 & .3 & .2 \\ -.1 & .8 & .3 \\ 0 & .5 & .5 \end{pmatrix}$

(d) $\mathbf{P} = \begin{pmatrix} .1 & .1 & .8 \\ 0 & 1 & 0 \\ .3 & 0 & .7 \end{pmatrix}$

2. Sue is a creature of habit. She eats only hamburgers or chicken salad for lunch. If she ate a hamburger yesterday, the probability is 0.8 that she will have chicken salad today. If she had chicken salad yesterday, there is a 50% probability that she will eat a hamburger today. Monday she had a hamburger. Find the probability that she will have a hamburger Wednesday by
   (a) drawing a tree diagram starting from Monday;
   (b) expressing Sue's luncheon habits as a Markov transition matrix; computing the probabilities of her Tuesday lunch choice; and using that to compute her Wednesday probabilities;
   (c) finding the two-step transition matrix and using it to compute Wednesday's probabilities directly.

3. Debby always lunches with Sue. Unlike Sue, she sometimes eats tuna fish. Letting s_1 = hamburger, s_2 = chicken salad and s_3 = tuna fish, the transition probabilities for Debby's lunch are given by

$$\mathbf{P} = \begin{pmatrix} .2 & .6 & .2 \\ .5 & 0 & .5 \\ .7 & .3 & 0 \end{pmatrix}$$

   If Debby flips a coin to decide between chicken salad and tuna fish on Wednesday, find the probability that she will eat fish on Friday. Find the probabilities of each state on Saturday.

4. Jo, Ed and Howie share the job of bringing the boss bad news. No one goes twice in a row. If Howie went last, Ed is a gentleman and goes twice as often as Jo. Howie is cowardly, so if Ed went last, Jo

goes three times as often as Howie. When the choice is between Howie and Ed, each goes equally often.

(a) Express this situation as a Markov process and write the transition matrix.

(b) The first time bad news came in, the three of them drew straws to see who would go. Find the probabilities of who went on the next occasion.

(c) Find the probabilities on the fourth occasion.

5. Returning to Exercise 2, the probability of hamburger and chicken salad on Monday is (5/13, 8/13) for Sue. What are the probabilities on Tuesday? Wednesday? Three weeks from Sunday?

6. Students at Baruch College hate math but never give up. Each student must complete Math 1 and Math 2 but 1/3 fail each course. Those who fail Math 1 repeat it; those who pass take Math 2. Those who fail Math 2 repeat it; those who pass go into a state of taking no Math. Express this situation as a Markov process. If 100% of all entering freshmen are enrolled in Math 1, what percent are still taking Math in their fourth semester?

7. In a town 40% of the workers are laborers and 60% are professionals. One-half of the laborers' sons are laborers and the remainder become professionals. It is three times as likely that the son of a professional will be a professional than a laborer. After two generations, what portion of the townsfolk are professionals?

8. There are four soap companies competing for a share of the market. Right now, Brand A has 50% of the market, Brand B has 30% and Brand C and D split the remainder equally. Each week people may change brands with the following type of behavior: Brand A retains 10% of its users and the rest divide equally among the other brands. Brand B retains none of its users who divide equally between A and C. Brand C and D users divide equally among all four brands. Write the transition matrix for this process and find the market distribution after two weeks.

9. The cigarette company that manufactures Brand C starts an aggressive advertising campaign. The results of this campaign are such that of people smoking Brand C in a given week, 80% continue to smoke it the following week; of those not smoking Brand C in a given week, 40% are won over to Brand C.

(a) Determine the transition matrix representing the above situation.

(b) If the probability that any person chosen at random smokes

Brand C today is 1/10, what is the probability that this person will smoke Brand C one week from now? Two weeks from now?

10. A final exam is given each term in finite math. The final exam is either very easy or slightly difficult (but not both). The probability that if it is easy this term, it will be easy next term is 0.4, whereas if it was slightly difficult one term, then it will be very easy the following term with probability 0.8.
    (a) Determine a transition matrix describing the above process.
    (b) Find the probability that the exam will be very easy next term if it is very easy this term.
    (c) If the probability that the exam will be easy this term is 1/4, what is the probability it will be very easy next term?
11. (a–e) Find $\mathbf{P}^3$ for the transition matrix $\mathbf{P}$ in Exercises 6 through 10.

5.6 Regular Markov Chains

In Exercise 5 of the last section, we saw an interesting example. In that case, we found that if $\mathbf{v}_0 = (5/13, 8/13)$, then $\mathbf{v}_1 = \mathbf{v}_0$ and hence $\mathbf{v}_2 = \mathbf{v}_1 = \mathbf{v}_0$ or, in general, $\mathbf{v}_n = \mathbf{v}_0$. That is, the probabilities of the two states remain constant from step to step. This type of behavior is called *equilibrium*. A good question now is, How does one find the equilibrium position of a Markov process? The answer is surprisingly simple. If $\mathbf{w}$ is the initial vector of probabilities and if after one step the vector of probabilities is again $\mathbf{w}$, then $\mathbf{wP} = \mathbf{w}$. Thus, suppose

$$\mathbf{P} = \begin{pmatrix} .6 & .4 \\ .8 & .2 \end{pmatrix}$$

Let $\mathbf{w} = (w_1, w_2)$ and try to solve

$$\mathbf{wP} = \mathbf{w}$$

We have

$$(w_1, w_2)\begin{pmatrix} .6 & .4 \\ .8 & .2 \end{pmatrix} = (w_1, w_2)$$

or

$$\begin{aligned}(0.6)w_1 + (0.8)w_2 &= w_1 \\ (0.4)w_1 + (0.2)w_2 &= w_2\end{aligned}$$

It pays to multiply both equations by 10 to eliminate decimals, leaving

$$6w_1 + 8w_2 = 10w_1$$
$$4w_1 + 2w_2 = 10w_2$$

The first equation is simply

$$8w_2 = 4w_1$$

and the second equation is the same! So we really have one equation in two unknowns. However, we also know that $w_1 + w_2 = 1$, since the w's are probabilities. Substituting $w_2 = 1 - w_1$ yields

$$8(1 - w_1) = 4w_1$$
$$8 = 12w_1$$
$$w_1 = 2/3 \qquad w_2 = 1/3$$

Hence, $\mathbf{w} = (2/3, 1/3)$ is the vector of equilibrium probabilities.

Let us look at another aspect of this equilibrium vector. Using, as before,

$$\mathbf{P} = \begin{pmatrix} .6 & .4 \\ .8 & .2 \end{pmatrix}$$

$$\mathbf{P}^2 = \begin{pmatrix} .6 & .4 \\ .8 & .2 \end{pmatrix}\begin{pmatrix} .6 & .4 \\ .8 & .2 \end{pmatrix} = \begin{pmatrix} .68 & .32 \\ .64 & .36 \end{pmatrix}$$

$$\mathbf{P}^4 = \begin{pmatrix} .68 & .32 \\ .64 & .36 \end{pmatrix}\begin{pmatrix} .68 & .32 \\ .64 & .36 \end{pmatrix} = \begin{pmatrix} .6672 & .3328 \\ .6656 & .3344 \end{pmatrix}$$

$$\mathbf{P}^8 = \begin{pmatrix} .6672 & .3328 \\ .6656 & .3344 \end{pmatrix}\begin{pmatrix} .6672 & .3328 \\ .6656 & .3344 \end{pmatrix}$$

$$= \begin{pmatrix} .66666752 & .33333248 \\ .66666596 & .33333504 \end{pmatrix}$$

It should not be hard to convince yourself that as n gets very large, $\mathbf{P}^n$ approaches

$$\mathbf{Q} = \begin{pmatrix} 2/3 & 1/3 \\ 2/3 & 1/3 \end{pmatrix}$$

That is, the rows of $\mathbf{P}^n$ both approach $\mathbf{w}$, the equilibrium vector. Moreover, let $\mathbf{v}_0 = (a, 1 - a)$ be the initial vector, and

$$\begin{aligned}
\mathbf{v}_0\mathbf{Q} &= (a, 1-a)\begin{pmatrix} 2/3 & 1/3 \\ 2/3 & 1/3 \end{pmatrix} \\
&= [(2/3)\,a + (2/3)\,(1-a), (1/3)\,a + (1/3)\,(1-a)] \\
&= (2/3, 1/3)
\end{aligned}$$

That is, no matter what the initial probability vector, after a large number of steps, the process will approach its equilibrium position! Now, this is not true for all Markov processes. To take an extreme example, let

$$\mathbf{P} = \begin{pmatrix} 0 & 1 \\ 1 & 0 \end{pmatrix}$$

Now, $\mathbf{w} = (1/2, 1/2)$ is the equilibrium vector. However, let $\mathbf{v}_0 = (1/3, 2/3)$. Then, $\mathbf{v}_1 = (2/3, 1/3)$; $\mathbf{v}_2 = (1/3, 2/3)$; $\mathbf{v}_3 = (2/3, 1/3)$ and so on. The process oscillates back and forth, never settling down toward the equilibrium status. However, most processes do tend to stabilize. In particular, we have the following definition.

Definition 6 Let $\mathbf{P}$ be a $(k \times k)$ transition matrix. If for some number n, $\mathbf{P}^n$ has no zero entries, then the process is called *regular*. For a regular process, it can be shown that

1. There exists a probability vector $\mathbf{w} = (w_1, w_2, \ldots, w_k)$ such that $\mathbf{wP} = \mathbf{w}$.
2. The rows of $\mathbf{P}^n$ all approach $\mathbf{w}$ as n gets large.
3. $\mathbf{v}_0\mathbf{P}^n$ approaches $\mathbf{w}$ regardless of $\mathbf{v}_0$.

Let us take one more example. There are three competing brands of cigars on the market. The most expensive is Swankismoke; the middle-priced brand is called Puffo; and the cheapest is called El Ropo. Each week, 20% of the El Ropo smokers buy Puffo instead. The rest buy El Ropo again. Twenty percent of the Puffo smokers buy El Ropo and 30% buy Swankismoke. The rest buy Puffo. Sixty percent of the Swankismoke users switch to Puffo and the rest are loyal to Swankismoke. Letting s_1 = Swankismoke, s_2 = Puffo and s_3 = El Ropo, the transition matrix is:

$$\mathbf{P} = \begin{pmatrix} .4 & .6 & 0 \\ .3 & .5 & .2 \\ 0 & .2 & .8 \end{pmatrix}$$

Now $\mathbf{P}^2 = \mathbf{PP}$

$$\mathbf{P}^2 = \begin{pmatrix} .4 & .6 & 0 \\ .3 & .5 & .2 \\ 0 & .2 & .8 \end{pmatrix}\begin{pmatrix} .4 & .6 & 0 \\ .3 & .5 & .2 \\ 0 & .2 & .8 \end{pmatrix}$$

$$= \begin{pmatrix} .34 & .54 & .12 \\ .27 & .47 & .26 \\ .06 & .26 & .68 \end{pmatrix}$$

Since $\mathbf{P}^2$ has no zero entries, $\mathbf{P}$ is a regular transition matrix. The equilibrium vector

$$\mathbf{w} = (w_1, w_2, w_3)$$

is found by solving

$$\mathbf{wP} = \mathbf{w}$$

$$(w_1, w_2, w_3)\begin{pmatrix} .4 & .6 & 0 \\ .3 & .5 & .2 \\ 0 & .2 & .8 \end{pmatrix} = (w_1, w_2, w_3)$$

$$\begin{aligned} 0.4w_1 + 0.3w_2 \qquad\qquad &= w_1 \\ 0.6w_1 + 0.5w_2 + 0.2w_3 &= w_2 \\ 0.2w_2 + 0.8w_3 &= w_3 \end{aligned}$$

Again multiplying through by 10, we have

$$\begin{aligned} 4w_1 + 3w_2 \qquad\quad &= 10w_1 \\ 6w_1 + 5w_2 + 2w_3 &= 10w_2 \\ 2w_2 + 8w_3 &= 10w_3 \\ -6w_1 + 3w_2 \qquad\quad &= 0 \\ 6w_1 - 5w_2 + 2w_3 &= 0 \\ 2w_2 - 2w_3 &= 0 \end{aligned}$$

This really gives us four equations in three unknowns, since

$$w_1 + w_2 + w_3 = 1$$

Writing the four equations in matrix form, we proceed to our usual triangularization:

$$\begin{pmatrix} -6 & 3 & 0 & \vdots & 0 \\ 6 & -5 & 2 & \vdots & 0 \\ 0 & 2 & -2 & \vdots & 0 \\ 1 & 1 & 1 & \vdots & 1 \end{pmatrix} \sim \begin{pmatrix} 1 & -1/2 & 0 & \vdots & 0 \\ 6 & -5 & 2 & \vdots & 0 \\ 0 & 2 & -2 & \vdots & 0 \\ 1 & 1 & 1 & \vdots & 1 \end{pmatrix}$$

$$\sim \begin{pmatrix} 1 & -1/2 & 0 & \vdots & 0 \\ 0 & -2 & 2 & \vdots & 0 \\ 0 & 2 & -2 & \vdots & 0 \\ 0 & 3/2 & 1 & \vdots & 1 \end{pmatrix} \sim \begin{pmatrix} 1 & -1/2 & 0 & \vdots & 0 \\ 0 & 1 & -1 & \vdots & 0 \\ 0 & 1 & -1 & \vdots & 0 \\ 0 & 3/2 & 1 & \vdots & 1 \end{pmatrix}$$

$$\sim \begin{pmatrix} 1 & -1/2 & 0 & \vdots & 0 \\ 0 & 1 & -1 & \vdots & 0 \\ 0 & 0 & 0 & \vdots & 0 \\ 0 & 0 & 5/2 & \vdots & 1 \end{pmatrix}$$

Back substituting yields $(5/2)w_3 = 1$, $w_3 = 2/5$; $w_2 - w_3 = 0$; $w_2 = 2/5$; and $w_1 - (1/2)w_2 = 0$; $w_1 = 1/5$. Thus, $\mathbf{w} = (0.2, 0.4, 0.4)$. That is, regardless of the original shares of the market, in the long run 20% of the smokers will buy Swankismoke and the remaining 80% of the market will be divided evenly between Puffo and El Ropo!

EXERCISES

1. Referring to Exercise 1, Section 5.5, find the equilibrium vectors for each transition matrix. Which matrices are regular?
2. Referring to Exercise 3, Section 5.5, find the percentage of the time Debby will eat tuna fish, in the long run.
3. Find the equilibrium probabilities for Jo, Ed and Howie in Exercise 4, Section 5.5.
4. Referring to Exercise 6, Section 5.5, find the equilibrium position for this process. Is this a regular process? Find $\mathbf{P}^2$, $\mathbf{P}^4$ and $\mathbf{P}^8$. Does the process appear to approach its equilibrium position?

5–8. Find the long-run state probabilities for the processes in Exercises 7–10, Section 5.5.

9. Given

$$\mathbf{P} = \begin{pmatrix} 1/2 & 1/2 & 0 & 0 \\ 0 & 1/3 & 2/3 & 0 \\ 0 & 0 & 1/3 & 2/3 \\ 1 & 0 & 0 & 0 \end{pmatrix}$$

Is $\mathbf{P}$ regular? Find its equilibrium vector.

10. The answer to a question is either yes or no. The information is passed from person to person in a chain of command. If at each step, there is a small probability p that the next person will misunderstand, find the transition matrix for this situation considered as a Markov process. Show that if the chain is very long, the probability that the information being passed is accurate will be 1/2.

5.7 Matrix Algebra

When we defined the product of two matrices, we said that the entry in Row i, Column j of **AB** would be the product of the ith row of **A** with the jth column of **B**. In order for this definition to hold, the rows of **A** must have as many components as the columns of **B**. If **A** is $m \times n$, then the number of components in each row of **A** is n. Similarly, if **B** is $p \times q$, then the number of components in each column of **B** is p. Thus, in order to form the product **AB**, we must have $n = p$. For example, let

$$\mathbf{A} = \begin{pmatrix} 2 & 1 & 3 & 4 \\ -1 & 3 & 0 & 2 \\ 1 & -1 & 2 & -2 \end{pmatrix} \quad \text{and} \quad \mathbf{B} = \begin{pmatrix} 1 & -1 \\ 2 & 0 \\ 1 & 1 \\ 3 & 5 \end{pmatrix}$$

Here, **A** is 3×4 and **B** is 4×2. Hence, we can form the product **AB** but *not* the product **BA**! The resulting matrix **AB** will have three rows and two columns. In fact

$$\mathbf{AB} = \begin{pmatrix} 2 & 1 & 3 & 4 \\ -1 & 3 & 0 & 2 \\ 1 & -1 & 2 & -2 \end{pmatrix} \begin{pmatrix} 1 & -1 \\ 2 & 0 \\ 1 & 1 \\ 3 & 5 \end{pmatrix} = \begin{pmatrix} 19 & 21 \\ 11 & 11 \\ -5 & -9 \end{pmatrix}$$

By the way, the simplest way to perform this multiplication is to start by multiplying Row 1 of **A** by each column of **B** until you run out of columns. The resulting products form the first row of **AB**. Now do it again, using the second row of **A** to form the second row of **AB**. Keep going until you run out of rows. You are now finished, and the number of rows and columns will come out straight automatically. However, just as a check, remember if **A** is $\underline{m} \times n$ and **B** is $n \times \underline{p}$, then **AB** will be $\underline{m} \times \underline{p}$.

Let us try an application. Below we have three tables. Table 1 gives the price per gross for stationery supplies. Table 2 gives the number of gross of each item shipped to each store in certain chains. Table 3 gives the number of chain stores in each of several cities.

Table 1

	Pencils	Pads	Erasers
Price	3.00	4.00	2.00
Shipping	.50	1.00	.50

Table 2

	Barbix Stores	Milban Stores	Newhill Stores	Coldel Stores
Pencils	2	4	1	6
Pads	2	0	1	3
Erasers	2	6	1	0

Table 3

	New York	Chicago	Boston
Barbix	1	2	1
Milban	0	3	2
Newhill	6	1	0
Coldel	2	2	2

Calling the cost matrix **A**, the number shipped per store matrix **B**, and the stores per city matrix **C**, we have **A** is 2×3, **B** is 3×4 and **C** is 4×3. We may form

$$\mathbf{AB} = \begin{pmatrix} 3 & 4 & 2 \\ .5 & 1 & .5 \end{pmatrix} \begin{pmatrix} 2 & 4 & 1 & 6 \\ 2 & 0 & 1 & 3 \\ 2 & 6 & 1 & 0 \end{pmatrix} = \begin{pmatrix} 18 & 24 & 9 & 30 \\ 4 & 5 & 2 & 6 \end{pmatrix}$$

Here we have multiplied cost per item by number of items per store, yielding cost to each store. Thus, for example, each Milban store (Column 2) is billed \$24 for stationery and \$5 for shipping. If we then multiply this by **C**, forming **(AB)C**, we get the total cost billed in each city. That is,

$$(\mathbf{AB})\mathbf{C} = \begin{pmatrix} 18 & 24 & 9 & 30 \\ 4 & 5 & 2 & 6 \end{pmatrix} \begin{pmatrix} 1 & 2 & 1 \\ 0 & 3 & 2 \\ 6 & 1 & 0 \\ 2 & 2 & 2 \end{pmatrix}$$

$$= \begin{pmatrix} 132 & 177 & 124 \\ 28 & 37 & 26 \end{pmatrix}$$

If, on the other hand, we had first formed **BC**, getting the amount shipped to each city, and then formed **A(BC)**, we should also end up with the total cost billed in each city. That is, **(AB)C** = **A(BC)**. In general, this associative law will hold. That is, **(AB)C** = **A(BC)** if all the products are defined. Hence, we can drop the parentheses and simply write **ABC**.

Please notice that we are very careful to maintain the order of multiplication. Thus, **(AB)C** = **A(BC)**, but we cannot write **AB** = **BA**! There are several reasons for this. First, it is entirely possible that **AB** is defined but **BA** is not. In fact, this is the case in the last example. But consider **B** and **C** in the last example. We could form **BC** and, having a 3×4 multiplying a 4×3, end up with a 3×3 matrix. The product **CB** could also be formed but the result would be 4×4. Thus **CB** ≠ **BC**. However, suppose we have square matrices. What then? Well, the product of two $(m \times m)$'s will be $m \times m$ no matter which is put first. Let us try

$$\mathbf{E} = \begin{pmatrix} 1 & 1 \\ 2 & 1 \end{pmatrix} \quad \text{and} \quad \mathbf{F} = \begin{pmatrix} -1 & 2 \\ 0 & 3 \end{pmatrix}$$

Now

$$\mathbf{EF} = \begin{pmatrix} -1 & 5 \\ -2 & 7 \end{pmatrix} \quad \text{and} \quad \mathbf{FE} = \begin{pmatrix} 3 & 1 \\ 6 & 3 \end{pmatrix}$$

Again, **EF** ≠ **FE**. One of the few times that **AB** = **BA** occurs is when **A** and **B** are powers of the same matrix. In fact, it is only this behavior that allows us to define $\mathbf{P}^n$ unambiguously.

We see that multiplication of a vector and a matrix is simply a special case of matrix multiplication. If a computer has been programmed to multiply matrices, it can handle a variety of bookkeeping operations in this context. For example, if we multiply a matrix **A** by a vector **U** = (1, 1, . . . , 1) with a suitable number of components, the resulting vector **UA** has as components the column sums of **A**. Thus, using **A** in the preceding example,

$$(1, 1)\begin{pmatrix} 3 & 4 & 2 \\ .5 & 1 & .5 \end{pmatrix} = (3.5, 5, 2.5)$$

we have a vector of total cost including shipping for each item. Multiplying a matrix **D** by **V**, a column vector of all ones, will yield for **DV** a vector of row sums. Thus, in the preceding example, **BC** is the matrix giving the numbers of pencils, pads and erasers shipped to New York, Chicago and Boston. The vector **(BC)V** gives the grand total of pencils shipped, pads shipped and erasers shipped.

$$\begin{pmatrix} 2 & 4 & 1 & 6 \\ 2 & 0 & 1 & 3 \\ 2 & 6 & 1 & 0 \end{pmatrix}\begin{pmatrix} 1 & 2 & 1 \\ 0 & 3 & 2 \\ 6 & 1 & 0 \\ 2 & 2 & 2 \end{pmatrix}\begin{pmatrix} 1 \\ 1 \\ 1 \end{pmatrix} = (71, 33, 47)$$

We see that 71 gross of pencils, 33 gross of pads and 47 gross of erasers have been shipped.

It may seem odd that we have defined multiplication of matrices but not addition. Let us look at a case in which addition might be called for. Suppose the matrix **B** gives the store orders for the month of January. Now, suppose that in February the orders were

$$\mathbf{D} = \begin{pmatrix} 2 & 1 & 1 & 0 \\ 1 & 4 & 2 & 0 \\ 1 & 1 & 3 & 4 \end{pmatrix}$$

The total orders for the two months would be a matrix, each of whose entries is the sum of the corresponding entries in **B** and **D**. We define this to be the sum of the two matrices. That is,

$$\mathbf{B} + \mathbf{D} = \begin{pmatrix} 2 & 4 & 1 & 6 \\ 2 & 0 & 1 & 3 \\ 2 & 6 & 1 & 0 \end{pmatrix} + \begin{pmatrix} 2 & 1 & 1 & 0 \\ 1 & 4 & 2 & 0 \\ 1 & 1 & 3 & 4 \end{pmatrix}$$

$$= \begin{pmatrix} 4 & 5 & 2 & 6 \\ 3 & 4 & 3 & 3 \\ 3 & 7 & 4 & 4 \end{pmatrix}$$

So we can add two matrices if they have the same numbers of rows and columns by simply adding entries in like positions. With such a definition, it is obvious that

$$\mathbf{A} + \mathbf{B} = \mathbf{B} + \mathbf{A} \tag{1}$$

$$(\mathbf{A} + \mathbf{B}) + \mathbf{C} = \mathbf{A} + (\mathbf{B} + \mathbf{C}) \tag{2}$$

It can be easily shown that

$$(\mathbf{A} + \mathbf{B})\mathbf{C} = \mathbf{AC} + \mathbf{BC} \qquad \mathbf{D}(\mathbf{A} + \mathbf{B}) = \mathbf{DA} + \mathbf{DB} \tag{3}$$

That is, multiplication is distributive over addition.

Now, if one were to take **A** + **A**, the results would be a matrix, each of whose entries is twice the corresponding entry in **A**. It would be nice to write **A** + **A** = 2**A**. In order to do so, one must define the product of a number k and a matrix **A** as follows: Let the entry in Row i, Column j of **A** be a_{ij}. The corresponding entry in k**A** would be ka_{ij}. In particular, if one forms the matrix (-1)**A**, then **A** + (-1)**A** will be a matrix with all zero entries. We shall call this a zero matrix and designate it **Z**. If any $m \times n$ matrix **A** is added to an $m \times n$ zero matrix, we have **A** + **Z** = **A**. Thus, the zero matrix acts just like the number zero.

If, as with numbers, we define $-$**A** to be the matrix which when added to **A** gives **Z** (that is, $-$**A** + **A** = **Z**), then it is easy to see that $-$**A** = (-1)**A** and we can add and subtract matrices just like numbers. For example, solve for x, y, z and w:

$$\begin{pmatrix} 1 & -x \\ 0 & 2 \end{pmatrix} + 2\begin{pmatrix} 1 & -1 \\ 2 & 1 \end{pmatrix} - 3\begin{pmatrix} 1 & -1 \\ y & 6 \end{pmatrix} = \begin{pmatrix} z & 0 \\ 11 & w \end{pmatrix}$$

Performing the indicated operations on the lefthand side yields

$$\begin{pmatrix} 0 & -x+1 \\ 4-3y & -14 \end{pmatrix} = \begin{pmatrix} z & 0 \\ 11 & w \end{pmatrix}$$

Now equating like elements, we have

$$\begin{aligned} z &= 0 & & \\ -x+1 &= 0 & x &= 1 \\ 4-3y &= 11 & y &= -7/3 \\ w &= -14 & & \end{aligned}$$

Interestingly, a zero matrix used in multiplication works like the number zero. Any matrix multiplied by **Z** gives **Z**(**AZ** = **Z**, **ZA** = **Z**). Of course, in such a statement it may be necessary to have two different meanings for **Z**. Thus, if **A** is 2×3 and **Z** is a 3×4 zero matrix, then **AZ** is a $\underline{2} \times 4$ zero matrix.

The major difference between the zero matrix and the number zero is that it is possible that the product **AB** is zero without either one being zero. For example, suppose

$$\mathbf{A} = \begin{pmatrix} 1 & -1 \\ -1 & 1 \end{pmatrix} \quad \text{and} \quad \mathbf{B} = \begin{pmatrix} 1 & 1 \\ 1 & 1 \end{pmatrix}$$

Now $\mathbf{AB} = \begin{pmatrix} 0 & 0 \\ 0 & 0 \end{pmatrix} = \mathbf{BA}$. Here both the products **AB** and **BA** are zero but neither is a zero matrix!

EXERCISES

1. Given $\mathbf{A} = \begin{pmatrix} -1 & 0 \\ 1 & 2 \\ 3 & 1 \\ -4 & -1 \end{pmatrix}$, $\mathbf{B} = \begin{pmatrix} 2 & 1 \\ -1 & 3 \\ -3 & 0 \\ 5 & 8 \end{pmatrix}$ and $\mathbf{C} = \begin{pmatrix} 1 & 2 \\ -3 & 5 \end{pmatrix}$,

   compute whichever of the following are well defined:
   (a) $\mathbf{AB} + \mathbf{C}$
   (b) $(\mathbf{A} + \mathbf{B})\mathbf{C}$
   (c) $\mathbf{A} + \mathbf{BC}$
   (d) $\mathbf{AC} + \mathbf{B}$
   (e) $(2\mathbf{A} - 3\mathbf{B})\mathbf{C}$
   (f) $\mathbf{C}(\mathbf{B} - \mathbf{A})$

2. Given $\mathbf{D} = \begin{pmatrix} 0 & 1 \\ -2 & 3 \end{pmatrix}$, $\mathbf{E} = \begin{pmatrix} 2 & -1 \\ 1 & -2 \end{pmatrix}$ and $\mathbf{F} = \begin{pmatrix} -3 & 2 \\ 1 & 4 \end{pmatrix}$,
   (a) verify $(\mathbf{DE})\mathbf{F} = \mathbf{D}(\mathbf{EF})$;
   (b) verify $\mathbf{D}(\mathbf{E} + \mathbf{F}) = \mathbf{DE} + \mathbf{DF}$;
   (c) show that $(\mathbf{D} + \mathbf{E})(\mathbf{D} + \mathbf{E}) \neq \mathbf{D}^2 + 2\mathbf{DE} + \mathbf{E}^2$. What does $(\mathbf{D} + \mathbf{E})(\mathbf{D} + \mathbf{E})$ equal?

3. Given $\mathbf{A} = \begin{pmatrix} 1 & 2 \\ -1 & -2 \end{pmatrix}$ and $\mathbf{B} = \begin{pmatrix} 2 & 4 \\ -1 & -2 \end{pmatrix}$, show that $\mathbf{AB} = \mathbf{Z}$ but $\mathbf{BA} \neq \mathbf{Z}$.

4. Given $\mathbf{A} = \begin{pmatrix} 2 & -3 \\ -2 & 3 \end{pmatrix}$, find a matrix $\mathbf{B}$ for which $\mathbf{AB} = \mathbf{Z}$.

5. Solve for x:

$$(x, 2)\left[\begin{pmatrix} 2 & 3 \\ -5 & 4 \end{pmatrix} - 2\begin{pmatrix} 1 & 2 \\ -3 & -1 \end{pmatrix}\right]\begin{pmatrix} 3 \\ x \end{pmatrix} = 33$$

6. Solve for a and b, if possible:

$$a\begin{pmatrix} 1 & 0 \\ 3 & 5 \end{pmatrix} + b\begin{pmatrix} 2 & 0 \\ 1 & 4 \end{pmatrix} = \begin{pmatrix} 1 & 0 \\ 3 & 2 \end{pmatrix}$$

7. For $\mathbf{A}, \mathbf{B}, \mathbf{C}, \mathbf{D}, \mathbf{E}, \mathbf{F}$ defined in Exercises 1 and 2, compute
   (a) $\mathbf{A}(\mathbf{D} + 3\mathbf{E} - 2\mathbf{F})$;
   (b) $(\mathbf{A} + \mathbf{B})(\mathbf{F} + 2\mathbf{C})$.

8. Compute $\mathbf{GH}$ and $\mathbf{HG}$ where $\mathbf{G} = (-1, 2, 4)$ and $\mathbf{H} = \begin{pmatrix} 1 \\ 3 \\ 5 \end{pmatrix}$.

9. Let $\mathbf{K} = \begin{pmatrix} 1 & 1 \\ 1 & 1 \end{pmatrix}$. Show that $\mathbf{K}^2 - 2\mathbf{K} = \mathbf{Z}$.

10. Table 1 shows the vitamin contents per gram in three diet supplements R, S and T. Table 2 shows the number of grams of the diet supplements in two different mixtures, M_1, M_2, of these supplements. Compute the table which shows the units of the four vitamins in the two mixtures.

Table 1

	R	S	T
Vitamin A	1	0	2
B	11	2	1
C	8	2	3
D	0	5	0

Table 2

	M_1	M_2
R	3	2
S	1	3
T	2	1

5.8 Identities and Inverses

For matrix addition, we have encountered a zero matrix **Z**, which plays the role of zero in addition. There also exists a class of matrices **I**, which play the role of one in multiplication, that is, $\mathbf{AI} = \mathbf{IA} = \mathbf{A}$. Again, we remark that if A is $m \times n$, then **I** will have to be $n \times n$ in the expression **AI**, and **I** will have to be $m \times m$ in the expression **IA**. The form of **I** is particularly simple. It is a square matrix with all zero entries, except along the diagonal that runs from top left to bottom right (called the principal diagonal). Along this diagonal the entries are all one. Thus,

$$\mathbf{I}^{(2\times 2)} = \begin{pmatrix} 1 & 0 \\ 0 & 1 \end{pmatrix} \qquad \mathbf{I}^{(3\times 3)} = \begin{pmatrix} 1 & 0 & 0 \\ 0 & 1 & 0 \\ 0 & 0 & 1 \end{pmatrix}$$

and so on.

The question is then, Does there exist an inverse matrix for any matrix **A**? That is, does there exist a matrix **B** such that $\mathbf{BA} = \mathbf{I}$? For example, for

$$\mathbf{A} = \begin{pmatrix} 0 & 0 \\ 1 & 2 \\ 1 & 1 \end{pmatrix} \qquad \mathbf{B} = \begin{pmatrix} 1 & -1 & 2 \\ 0 & 1 & -1 \end{pmatrix}$$

we have

$$\mathbf{BA} = \begin{pmatrix} 1 & -1 & 2 \\ 0 & 1 & -1 \end{pmatrix}\begin{pmatrix} 0 & 0 \\ 1 & 2 \\ 1 & 1 \end{pmatrix} = \begin{pmatrix} 1 & 0 \\ 0 & 1 \end{pmatrix} = \mathbf{I}$$

Thus, we would say **B** is a "left inverse" for **A**. Unfortunately,

$$\mathbf{AB} = \begin{pmatrix} 0 & 0 \\ 1 & 2 \\ 1 & 1 \end{pmatrix}\begin{pmatrix} 1 & -1 & 2 \\ 0 & 1 & -1 \end{pmatrix} = \begin{pmatrix} 0 & 0 & 0 \\ 1 & 1 & 0 \\ 1 & 0 & 1 \end{pmatrix} \neq \mathbf{I}$$

Hence, **B** is not a "right inverse" for **A**.

For $m \times n$ matrices, $m \neq n$, there is no way in which one can construct a single matrix **B**, so that $\mathbf{AB} = \mathbf{I}$ and $\mathbf{BA} = \mathbf{I}$. However, for square matrices the opposite is true. That is, if $\mathbf{AB} = \mathbf{I}$, then $\mathbf{BA} = \mathbf{I}$ and we can refer to **B** as the multiplicative inverse of **A**. We shall designate this matrix $\mathbf{A}^{-1}$ and call it "**A** inverse."

To find $\mathbf{A}^{-1}$, one could proceed directly from the definition. For example, let

$$\mathbf{A} = \begin{pmatrix} 2 & 1 \\ 1 & 1 \end{pmatrix}$$

We seek

$$\mathbf{A}^{-1} = \begin{pmatrix} a & b \\ c & d \end{pmatrix}$$

So that $\mathbf{AA}^{-1} = \mathbf{I}$. Thus,

$$\begin{pmatrix} 2 & 1 \\ 1 & 1 \end{pmatrix}\begin{pmatrix} a & b \\ c & d \end{pmatrix} = \begin{pmatrix} 1 & 0 \\ 0 & 1 \end{pmatrix}$$

This yields

$$\begin{aligned} 2a + c &= 1 \qquad & 2b + d &= 0 \\ a + c &= 0 \qquad & b + d &= 1 \\ a = 1, c &= -1 \qquad & b = -1, d &= 2 \end{aligned}$$

The result is

$$\mathbf{A}^{-1} = \begin{pmatrix} 1 & -1 \\ -1 & 2 \end{pmatrix}$$

It is wise to check this by showing that $\mathbf{A}^{-1}\mathbf{A} = \mathbf{I}$, thus verifying that although we solved for $\mathbf{A}^{-1}$ by seeking a right inverse, the resulting matrix is also the left inverse.

Unfortunately, not all matrices are invertible. For example, if one attempts to solve

$$\begin{pmatrix} 1 & 1 \\ 2 & 2 \end{pmatrix}\begin{pmatrix} a & b \\ c & d \end{pmatrix} = \begin{pmatrix} 1 & 0 \\ 0 & 1 \end{pmatrix}$$

the result is

$$\begin{aligned} a + c &= 1 \qquad & b + d &= 0 \\ 2a + 2c &= 0 \qquad & 2b + 2d &= 1 \end{aligned}$$

Both sets of equations are incompatible and hence $\begin{pmatrix} 1 & 1 \\ 2 & 2 \end{pmatrix}^{-1}$ does not exist.

Let us try a 3×3 matrix.

$$\mathbf{A} = \begin{pmatrix} 1 & -1 & 2 \\ 1 & -2 & 3 \\ 3 & -1 & 2 \end{pmatrix} \qquad \mathbf{A}^{-1} = \begin{pmatrix} a & d & g \\ b & e & h \\ c & f & i \end{pmatrix}$$

$$\mathbf{A}\mathbf{A}^{-1} = \begin{pmatrix} 1 & -1 & 2 \\ 1 & -2 & 3 \\ 3 & -1 & 2 \end{pmatrix}\begin{pmatrix} a & d & g \\ b & e & h \\ c & f & i \end{pmatrix} = \begin{pmatrix} 1 & 0 & 0 \\ 0 & 1 & 0 \\ 0 & 0 & 1 \end{pmatrix}$$

$$\begin{aligned} a - b + 2c &= 1 \qquad & d - e + 2f &= 0 \qquad & g - h + 2i &= 0 \\ a - 2b + 3f &= 0 \qquad & d - 2e + 3f &= 1 \qquad & g - 2h + 3i &= 0 \\ 3a - b + 2c &= 0 \qquad & 3d - e + 2f &= 0 \qquad & 3g - h + 2i &= 1 \end{aligned}$$

An inspection of the three sets of equations shows that if we write them in matrix form, the lefthand side is in every case the same. We wish to solve

$$\left(\begin{array}{ccc|c} 1 & -1 & 2 & 1 \\ 1 & -2 & 3 & 0 \\ 3 & -1 & 2 & 0 \end{array}\right) \qquad \left(\begin{array}{ccc|c} 1 & -1 & 2 & 0 \\ 1 & -2 & 3 & 1 \\ 3 & -1 & 2 & 0 \end{array}\right) \qquad \left(\begin{array}{ccc|c} 1 & -1 & 2 & 0 \\ 1 & -2 & 3 & 0 \\ 3 & -1 & 2 & 1 \end{array}\right)$$

Since the required row operations would be the same in each case, we may as well solve them all at once! Thus we write

$$\left(\begin{array}{ccc|ccc} 1 & -1 & 2 & 1 & 0 & 0 \\ 1 & -2 & 3 & 0 & 1 & 0 \\ 3 & -1 & 2 & 0 & 0 & 1 \end{array}\right)$$

Now, the first three columns are **A** and the last three columns are **I**. Let us start solving:

$$\left(\begin{array}{ccc|ccc} 1 & -1 & 2 & 1 & 0 & 0 \\ 1 & -2 & 3 & 0 & 1 & 0 \\ 3 & -1 & 2 & 0 & 0 & 1 \end{array}\right) \sim \left(\begin{array}{ccc|ccc} 1 & -1 & 2 & 1 & 0 & 0 \\ 0 & -1 & 1 & -1 & 1 & 0 \\ 0 & 2 & -4 & -3 & 0 & 1 \end{array}\right)$$

$$\left(\begin{array}{ccc|ccc} 1 & -1 & 2 & 1 & 0 & 0 \\ 0 & 1 & -1 & 1 & -1 & 0 \\ 0 & 0 & -2 & -5 & -2 & 1 \end{array}\right)$$

Having triangularized, let us interpret. Remember that the last three columns represent the righthand sides of each of the three systems in turn. So we have $-2c = -5$, $c = 2\frac{1}{2}$; $-2f = -2, f = 1$; and $-2i = 1$, $i = -1/2$ and we can now back substitute as in chapter 2. On the other hand, had we simply divided the bottom row by -2 (an allowable row operation), reducing the first half to (0, 0, 1), then the last three elements would form the bottom row of $\mathbf{A}^{-1}$. However, it is more convenient to reduce the first half of each row to (1, 0, 0), (0, 1, 0) and (0, 0, 1) (i.e., the rows of **I**). In that case, the last half will be the corresponding rows of $\mathbf{A}^{-1}$. Do you see why?

In other words, if we reduce **A** to **I** by row operations, we simultaneously reduce **I** to $\mathbf{A}^{-1}$. To accomplish this "total elimination," we proceed exactly as for triangularization, except instead of clearing those elements in each column below a specific row, we will clear all elements above and below! Thus,

$$(\mathbf{A} \mid \mathbf{I}) = \left(\begin{array}{ccc|ccc} 1 & -1 & 2 & 1 & 0 & 0 \\ 1 & -2 & 3 & 0 & 1 & 0 \\ 3 & -1 & 2 & 0 & 0 & 1 \end{array}\right)$$

$$\sim \left(\begin{array}{ccc|ccc} 1 & -1 & 2 & 1 & 0 & 0 \\ 0 & -1 & 1 & -1 & 1 & 0 \\ 0 & 2 & -4 & -3 & 0 & 1 \end{array}\right) \quad \begin{array}{l} \text{Row 1} \\ \text{Row 2} - \text{Row 1} \\ \text{Row 3} - 3 \times \text{Row 1} \end{array}$$

$$\sim \left(\begin{array}{ccc|ccc} 1 & 0 & 1 & 2 & -1 & 0 \\ 0 & 1 & -1 & 1 & -1 & 0 \\ 0 & 0 & -2 & -5 & 2 & 1 \end{array}\right) \quad \begin{array}{l} \text{Row 1} - \text{Row 2} \\ (-1)\ \text{Row 2} \\ \text{Row 3} + 2 \times \text{Row 2} \end{array}$$

$$\sim \left(\begin{array}{ccc|ccc} 1 & 0 & 0 & -1/2 & 0 & 1/2 \\ 0 & 1 & 0 & 7/2 & -2 & -1/2 \\ 0 & 0 & 1 & 5/2 & -1 & -1/2 \end{array}\right) \quad \begin{array}{l} \text{Row 1} + 1/2 \times \text{Row 3} \\ \text{Row 2} - 1/2 \times \text{Row 3} \\ -1/2 \times \text{Row 3} \end{array}$$

We now have $(\mathbf{I} \vdots \mathbf{A}^{-1})$, so

$$\mathbf{A}^{-1} = \begin{pmatrix} -1/2 & 0 & 1/2 \\ 7/2 & -2 & -1/2 \\ 5/2 & -1 & -1/2 \end{pmatrix} = 1/2 \begin{pmatrix} -1 & 0 & 1 \\ 7 & -4 & -1 \\ 5 & -2 & -1 \end{pmatrix}$$

Again, the student should check this result. Naturally, this same technique would be used for 2×2, 4×4 or any other square matrix.

Suppose one wished to solve the system

$$\begin{aligned} x - y + 2z &= 3 \\ x - 2y + 3z &= 4 \\ 3x - y + 2z &= 5 \end{aligned}$$

Let the matrix of coefficients be

$$\mathbf{A} = \begin{pmatrix} 1 & -1 & 2 \\ 1 & -2 & 3 \\ 3 & -1 & 2 \end{pmatrix}$$

Let

$$\mathbf{X} = \begin{pmatrix} x \\ y \\ z \end{pmatrix}$$

and let

$$\mathbf{B} = \begin{pmatrix} 3 \\ 4 \\ 5 \end{pmatrix}$$

We have the system written as

$$\mathbf{AX} = \mathbf{B}$$

Multiplying both sides of this equation by $\mathbf{A}^{-1}$, we have

$$\begin{aligned} \mathbf{A}^{-1}\mathbf{AX} &= \mathbf{A}^{-1}\mathbf{B} \\ \mathbf{IX} &= \mathbf{A}^{-1}\mathbf{B} \\ \mathbf{X} &= \mathbf{A}^{-1}\mathbf{B} \end{aligned}$$

Since we just computed $\mathbf{A}^{-1}$, we have, in other words,

$$\begin{pmatrix} x \\ y \\ z \end{pmatrix} = 1/2\begin{pmatrix} -1 & 0 & 1 \\ 7 & -4 & -1 \\ 5 & -2 & -1 \end{pmatrix}\begin{pmatrix} 3 \\ 4 \\ 5 \end{pmatrix} = 1/2\begin{pmatrix} 2 \\ 0 \\ 2 \end{pmatrix} = \begin{pmatrix} 1 \\ 0 \\ 1 \end{pmatrix}$$

Thus, $x = 1$, $y = 0$, $z = 1$ solves the system.

Does it pay to solve a system this way? Surely not, since the row operations needed to invert the matrix are more extensive than those needed to solve the system. Of course, if one had reason to need the solution to $\mathbf{AX} = \mathbf{B}$ for several different $\mathbf{B}$'s, then it might pay to do the operations once and compute $\mathbf{A}^{-1}$.

An interesting sidelight on the existence of inverses occurs in Markov processes. If $\mathbf{P}$ is a transition matrix and $\mathbf{w}$ is its equilibrium vector, then $\mathbf{wP} = \mathbf{w}$. In other words,

$$\mathbf{wP} - \mathbf{w} = \mathbf{Z}$$
$$\mathbf{w}(\mathbf{P} - \mathbf{I}) = \mathbf{Z}$$

If $(\mathbf{P} - \mathbf{I})^{-1}$ exists, then

$$\mathbf{w} = \mathbf{Z}(\mathbf{P} - \mathbf{I})^{-1} = \mathbf{Z}$$

Thus, the equilibrium vector would have to be zero. Since we have stated that every transition matrix has a nonzero equilibrium vector, it follows that $\mathbf{P} - \mathbf{I}$ has no inverse.

EXERCISES

1. Given $\mathbf{A} = \begin{pmatrix} -1 & 2 \\ 1 & 3 \end{pmatrix}$, $\mathbf{B} = \begin{pmatrix} -1 & 3 \\ 2 & -4 \end{pmatrix}$, find $\mathbf{AB}$, $(\mathbf{AB})^{-1}$, $\mathbf{A}^{-1}$ and $\mathbf{B}^{-1}$. Show that $\mathbf{B}^{-1}\mathbf{A}^{-1} = (\mathbf{AB})^{-1}$.
2. Show that if $\mathbf{A}$ and $\mathbf{B}$ are any two invertible matrices, $(\mathbf{AB})^{-1} = \mathbf{B}^{-1}\mathbf{A}^{-1}$.
3. Use $\mathbf{A}$ and $\mathbf{B}$ in Exercise 1 to show that $(\mathbf{A} + \mathbf{B})^{-1} \neq \mathbf{A}^{-1} + \mathbf{B}^{-1}$.
4. Given $\mathbf{A} = \begin{pmatrix} 1 & -1 \\ 2 & -3 \end{pmatrix}$, $\mathbf{B} = \begin{pmatrix} 4 & -1 & 2 \\ 2 & 1 & 3 \end{pmatrix}$ and $\mathbf{C} = \begin{pmatrix} 2 & 0 & 1 \\ -1 & 2 & 1 \end{pmatrix}$, find $\mathbf{A}^{-1}\mathbf{B} + 3\mathbf{C}$.
5. (a) Find $\mathbf{A}^{-1}$ if

$$\mathbf{A} = \begin{pmatrix} 1 & 2 & -3 \\ 2 & 3 & 4 \\ 3 & 2 & 5 \end{pmatrix}$$

(b) Solve the following system of equations:

$$\begin{aligned} x + 2y - 3z &= 6 \\ 2x + 3y + 4z &= 1 \\ 3x + 2y + 5z &= 0 \end{aligned}$$

6. (a) Find $\mathbf{A}^{-1}$ if

$$\mathbf{A} = \begin{pmatrix} 1 & 2 & -2 \\ 2 & 5 & 1 \\ -5 & -10 & 9 \end{pmatrix}$$

(b) Solve the following system of equations:

$$\begin{aligned} x + 2y - 2z &= -5 \\ 2x + 5y + z &= 12 \\ -5x - 10y + 9z &= 21 \end{aligned}$$

7. Simplify:

$$\begin{pmatrix} 1 & 3 \\ 0 & -2 \\ 2 & 4 \end{pmatrix}\begin{pmatrix} 3 & 3 \\ 2 & -2 \end{pmatrix}^{-1} - 3\begin{pmatrix} 2 & -1 \\ 2 & 3 \\ 3 & 0 \end{pmatrix}$$

8. Let $\mathbf{M}$ be the transition matrix for a Markov process, $\mathbf{M} = \begin{pmatrix} .2 & .8 \\ .4 & .6 \end{pmatrix}$. Show that $\mathbf{M} - \mathbf{I}$ has no inverse.

9. Find a matrix $\mathbf{B}$ such that $\mathbf{AB} = \mathbf{I}$, where $\mathbf{A} = \begin{pmatrix} 1 & 2 & 1 \\ -1 & 4 & 3 \end{pmatrix}$.

10. (a) Find $\mathbf{A}^{-1}$ if

$$\mathbf{A} = \begin{pmatrix} 1 & -1 & 3 & 2 \\ 2 & -2 & 2 & -1 \\ 3 & 0 & 0 & -2 \\ 4 & 3 & 1 & 1 \end{pmatrix}$$

(b) Find $\mathbf{A}^{-1}$ if

$$\mathbf{A} = \begin{pmatrix} 1 & 2 & -1 & 1 \\ -1 & -3 & 4 & 2 \\ 2 & 2 & 3 & -2 \\ 3 & 2 & 1 & 1 \end{pmatrix}$$

5.9 Linear Programming Revisited

As an application of the method of complete elimination, we shall consider a purely algebraic method of solving linear programming problems. The student should recall that by the enumeration of vertices one may solve a linear programming problem without resorting to the geometric construction. However, in a fairly large problem, the number of vertices soon becomes so enormous that a more efficient technique is required. The technique which we shall discuss is called the *simplex procedure*. The beauty of the simplex procedure is that one needs only to detect vertices that fall in the region of feasible solutions. Additionally, starting from one feasible vertex, the procedure will lead you to another feasible solution for which the objective function is larger, and it will tell you to stop when the optimal solution is reached. Finally, the steps are so routine and repetitious that the whole method is readily programmed for use by computers, thus making it applicable to large-scale problems.

To fix the ideas, let us take a specific example in standard form.

$$\begin{aligned} \text{Maximize:} \quad & P = x + 2y \\ \text{Subject to:} \quad & 3x + y \leq 15 \\ & 3x + 4y \leq 24 \\ & x \geq 0 \\ & y \geq 0 \end{aligned}$$

Notice that the standard form involves three basic requirements:

1. The objective is to be maximized.
2. The unknowns must be nonnegative.
3. The other constraints must be of the form

$$ax + by + \cdots + ew \leq k$$

for unknowns $x, y, \ldots, w$.

To start, figure 13 shows the geometric solution. We see that the objective P is maximized at $(0, 6)$ where $P = 12$. However, let us see how to solve the problem algebraically.

First, we convert the inequalities into equalities by introducing nonnegative slack variables r, s and we convert the objective into the equation

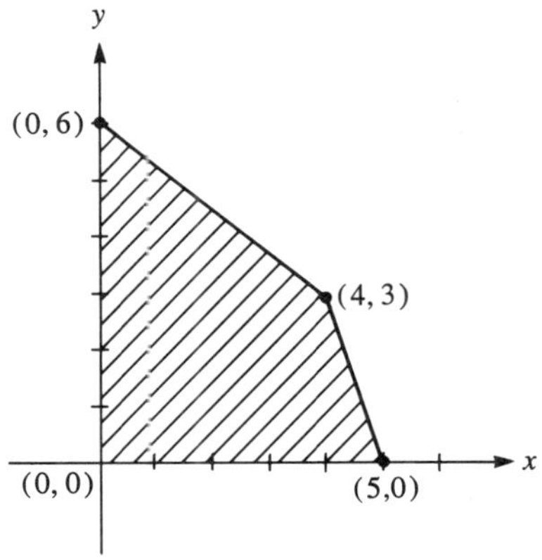

Figure 13

$-x - 2y + P = 0$. Thus,

$$\begin{aligned} 3x + y + r \quad\quad &= 15 \\ 3x + 4y \quad + s \quad &= 24 \\ -x - 2y \quad\quad + P &= 0 \end{aligned}$$

We rewrite this system in a special matrix form called the *simplex tableau.*

x	y	r	s	P	
3	1	1	0	0	15
3	4	0	1	0	24
−1	−2	0	0	1	0

This represents three equations in five unknowns, which has many solutions. We shall read only the simplest solution detectable by inspection, namely,

$$\begin{aligned} r &= 15 \\ s &= 24 \\ P &= 0 \\ x &= 0 \\ y &= 0 \end{aligned}$$

That is, if in a column we have several nonzero entries, then that unknown is assigned the value zero. Now, clearly the solution we have corresponds to the vertex (0, 0) and $P = 0$ which is surely not the optimal solution. Since the last row says $-x - 2y + P = 0$, if we increase either x or y we increase also P since the sum must remain zero. Which should we increase? Every unit increase in y causes two units increase in P. But every unit increase in x causes one unit increase in P. Hence, it seems reasonable to increase y. By how much should we increase y? The first equation says $3x + y + r = 15$. The largest possible value of y is attained at $x = r = 0$. Hence $y = 15$. However, if $y = 15$, the second equation becomes

$$\begin{aligned} 3x + 4(15) + s &= 24 \\ 3x + s &= -36 \end{aligned}$$

This means either x or s must be negative, which is not allowed. Hence, the maximum allowable value of y is attained when $x = s = 0$ in the second equation. That is,

$$4y = 24 \qquad y = 6$$

Now, in order to read $y = 6$ as part of the solution of the problem, we must perform a complete elimination clearing all of Column 2 except in Row 2. We shall call the element in Row 2, Column 2, the "pivotal element," which we circle thus:

x	y	r	s	P	
3	1	1	0	0	15
3	(4)	0	1	0	24
-1	-2	0	0	1	0

x	y	r	s	P		
11/4	0	1	$-3/4$	0	9	Row 1 − 1/4 × Row 2
3/4	1	0	1/4	0	6	Row 2 ÷ 4
1/2	0	0	1/2	1	12	Row 3 + 1/2 × Row 2

The solution reads $x = 0$, $y = 6$, $r = 9$, $s = 0$, $P = 12$. This corresponds to the vertex (0, 6), which is the optimal point. How do we know? The last equation reads

$$(1/2)x + (1/2)s + P = 12$$

Since the coefficients of x and s are positive, any increase in either causes P to become smaller. Hence, we must stop.

Let us review the method used. First, to find the pivotal element we identified the proper column by taking the one with the most negative entry at the bottom. Having picked the proper column, the proper row was found as follows: Take the ratio of the last element in each row to the element in the selected column. The row for which this ratio is the least is the proper one. In the last example, we chose the second column because its last entry was -2. The pertinent ratios were then 15/1 for the first row and 24/4 for the second row. Hence, the second row was chosen.

How do we know when to stop? When there are no more negative entries in the last row! For example, take the same problem, only make the objective function $P = 2x + y$. That is,

$$\begin{aligned} \text{Maximize:}\quad & P = 2x + y \\ \text{Subject to:}\quad & 3x + y \leq 15 \\ & 3x + 4y \leq 24 \\ & x \geq 0 \\ & y \geq 0 \end{aligned}$$

The tableau is

x	y	r	s	P	
(3)	1	1	0	0	15
3	4	0	1	0	24
-2	-1	0	0	1	0

The most negative entry, -2, is in the x-column. The corresponding ratios are 15/3 and 24/3. Hence, the 3 in the upper left-hand corner has been circled as the pivotal element. Carry out the total elimination process on the x-column:

x	y	r	s	P	
1	1/3	1/3	0	0	5
0	(3)	-1	1	0	9
0	$-1/3$	2/3	0	1	10

We still have a negative element in the y-column; hence P may be increased further by introducing y into the solution. We compute the pertinent ratios, which are $5 \div (1/3) = 15$ and $9/3 = 3$. The pivotal element is then the 3 circled above. Perform the total elimination on the y-column:

x	y	r	s	P	
1	0	4/9	−1/9	0	4
0	1	−1/3	1/3	0	3
0	0	5/9	1/9	1	11

We now have no negative entries in the last column. We stop the process and read the solution:

$$x = 4$$
$$y = 3$$
$$r = 0$$
$$s = 0$$
$$P = 11$$

It is easy to see geometrically that this is the optimal solution. Notice that in the first example, the final answer showed one of the slack variables r to equal 9. That is, there is some slack in the solution. In other words, if 15 and 24 represent the total quantities of certain resources, then in this solution something is left over at the optimal point. In the second example, both slack variables are zero, indicating that the best solution utilizes the total resources available.

It should be noted that the simplex procedure offers one other advantage. Since it is totally arithmetic, the technique will work with any number of variables. Let us try one somewhat larger problem.

The New York Nut Co. packages 8-ounce jars of nuts under three different labels, Econo-nut, All-nutz and Lux-O-nut. Each is a mixture of peanuts, cashews and almonds. In the following table, we have the breakdown and profit per jar for each of the three:

	Peanuts	Cashews	Almonds	Profit
Econo-nut	6 oz	1 oz	1 oz	10¢
All-nutz	4 "	3 "	1 "	20¢
Lux-O-nuts	2 "	2 "	4 "	30¢

Suppose the total number of ounces available of peanuts, cashews and almonds are 800, 600 and 300 ounces respectively. How many of each jar should be prepared in order to maximize the profit?

Let

x = number of jars of Econo-nut

y = number of jars of All-nutz

z = number of jars of Lux-O-nuts

Then the problem becomes

$$\begin{aligned} \text{Maximize: } P &= 10x + 20y + 30z \\ \text{Subject to: } 6x + 4y + 2z &\le 800 \\ x + 3y + 2z &\le 600 \\ x + y + 4z &\le 300 \\ x &\ge 0 \\ y &\ge 0 \\ z &\ge 0 \end{aligned}$$

We introduce three slack variables r, s, t and arrive at the simplex tableau:

x	y	z	r	s	t	P	
6	4	2	1	0	0	0	800
1	3	2	0	1	0	0	600
1	1	(4)	0	0	1	0	300
−10	−20	−30	0	0	0	1	0

We have circled the pivotal element in the above. After total elimination the tableau is

x	y	z	r	s	t	P	
5.5	3.5	0	1	0	−.5	0	650
.5	(2.5)	0	0	1	−.5	0	450
.25	.25	1	0	0	.25	0	75
−2.5	−12.5	0	0	0	7.5	1	2250

Again, we have circled the pivotal element. After total elimination, the tableau is

x	y	z	r	s	t	P	
4.8	0	0	1	−1.4	.9	0	20
.2	1	0	0	.4	−.2	0	180
.2	0	1	0	−.1	.3	0	30
0	0	0	0	4	5	1	4500

There are now no negative entries in the last row. Hence we read the solution: $x = 0$, $y = 180$, $z = 30$, $r = 20$, $s = 0$, $t = 0$, $P = 4500$. Thus, the company should sell 180 jars of All-nutz, 30 jars of Lux-O-nuts and no Econo-nut. The number $r = 20$ indicates that 20 ounces of peanuts will be left over. The optimal profit is $P = 4500¢ = \$45$ if this procedure is followed.

A few closing remarks are in order. Suppose when choosing the pivotal element two columns tie for having the most negative bottom entry. We may then choose either column first. Having chosen a column, we are supposed to select the pivotal element as the one for which a certain ratio is smallest. This is not quite accurate. If the smallest ratio is negative, this corresponds to a negative value for the unknown, which is not allowed in the set of feasible solutions. Hence, we should say, Choose the element for which one obtains the smallest nonnegative ratio. Again, if there is a tie for lowest in this case, choose the one you wish.

EXERCISES

1. Maximize: $x + 2y + 3z$

   Subject to:
   $$\begin{aligned} 2x + y + z &\le 2 \\ 4x + 2y + z &\le 2 \\ x, y, z &\ge 0 \end{aligned}$$

2. Maximize: $2w - x + 3y + z$

   Subject to:
   $$\begin{aligned} 3w - x - y + z &\le 4 \\ w - x + y + z &\le 1 \\ 2w + x + 2y &\le 7 \\ w, x, y, z &\ge 0 \end{aligned}$$

3. Maximize: $P = x + 3y$

   Subject to:
$$\begin{aligned} y &\le 5 \\ 3x + 5y &\le 34 \\ 2x + y &\le 18 \\ x &\ge 0 \\ y &\ge 0 \end{aligned}$$

4. A coffee manufacturer makes 50¢ profit on each can of instant coffee, and 35¢ profit on each can of regular coffee. The manufacturer must include for use in each can beans from both Brazil and Colombia. In his storeroom he has available 1,200 ounces of Brazilian beans and 1,500 ounces of Colombian beans. If each can of instant coffee requires 4 oz of Brazilian beans, and 3 oz of Colombian beans, while each can of regular coffee requires 2 oz of Brazilian beans and 3 oz of Colombian beans, how many cans of each type should the manufacturer produce to maximize his profit?

5. Maximize: $4x - 6y + 2z$

   Subject to:
$$\begin{aligned} 2x + 2y + 2z &\le 8 \\ x \qquad - z &\le 2 \\ x, y, z &\ge 0 \end{aligned}$$

6. Maximize: $4x + 3y + 7z$

   Subject to:
$$\begin{aligned} x + 2y + 3z &\le 100 \\ 3x + y + 3z &\le 100 \\ x, y, z &\ge 0 \end{aligned}$$

7. Suppose that in the New York Nut Co. example in the text, the company changes the contents of the Econo-nut jar by substituting an extra ounce of peanuts for the almonds. Suppose that by doing so the company increases the profit to 15¢/jar. How many of each type jar should they now market?

8. A television assembly plant divides its operation among three shops, electronic assembly, cabinet assembly, and packing and shipping. The plant assembles cheap floor models, portables and luxury floor models. The table below shows the man hours needed for each model, the number of hours available in each shop and the price on each type of model. Find the number of each type to be manufactured to maximize total sales.

	Electronic Assembly	Cabinet Assembly	Packing and Shipping	Price
Cheap Model	3	2	2	\$350
Portable Model	2	1	2	\$300
Deluxe Model	3	4	2	\$600
Hours Available	105	120	70	

5.10 Artificial Variables

In the last section we dealt with linear programming problems in standard form. However, in chapter 2, we solved many linear programming problems in which some or all of the inequalities were in nonstandard form or in which the objective was to be minimized rather than maximized. Take the following example:

$$\begin{aligned} \text{Minimize: } & C = x + 2y \\ \text{Subject to: } & x - y \leq 4 \\ & 2x + y \geq 4 \\ & 4x - y \leq 8 \\ & x, y \geq 0 \end{aligned}$$

We would naturally first introduce nonnegative slack variables to make the constraints into equalities. In the first and third inequalities, we must add something, but in the second, we must subtract a suitable slack variable, yielding

$$\begin{aligned} x - y + r \qquad\qquad &= 4 \\ 2x + y \qquad - s \qquad &= 4 \\ 4x - y \qquad\qquad + t &= 8 \\ x, y, r, s, t &\geq 0 \end{aligned}$$

Furthermore, we must adjust the objective so that the problem becomes a maximization problem. This is accomplished by letting $P = -C = -x - 2y$. Now maximizing P is the same as minimizing C. The objective rewritten in equation form is $x + 2y + P = 0$ and the simplex tableau is

x	y	r	s	t	P	
1	−1	1	0	0	0	4
2	1	0	−1	0	0	4
4	−1	0	0	1	0	8
1	2	0	0	0	1	0

This tableau has two drawbacks. First, the only solution which can be read off directly is $x = 0$, $y = 0$, $r = 4$, $s = -4$, $t = 8$ and $P = 0$. This is not a feasible solution since one variable has a negative value. Second, even if there were a feasible solution, we have no negative elements in the last row and hence no way to find a pivotal element and get started.

The first problem is avoided by introducing an "artificial" variable u into the second equation, making it

$$2x + y - s + u = 4$$

Of course, if two variables in the basic solution were negative, we would need two artificial variables and so on. In this case, the basic solution is now

$$x = y = s = P = 0 \qquad u = 4,\ r = 4,\ t = 8$$

We know that slack variables do not enter into the objective function but what about artificial variables? If we form a new objective $Q = P - nu$, where n is a large number, then it will be impossible to maximize Q unless $u = 0$ and P is maximized. Thus, the final form of our problem is

$$\begin{aligned} \text{Maximize: } & Q = -x - 2y - nu \\ \text{Subject to: } & x - y + r = 4 \\ & 2x + y - s + u = 4 \\ & 4x - y + t = 8 \\ & x, y, r, s, t, u \geq 0 \end{aligned}$$

and in tableau form we have

x	y	r	s	u	t	Q	
1	−1	1	0	0	0	0	4
2	1	0	−1	1	0	0	4
4	−1	0	0	0	1	0	8
1	2	0	0	n	0	1	0

What is lacking is a negative element in the last row and its associated pivotal element. In order to have an equivalent system of equations with negative elements in the last row, we will eliminate u from the fourth equation by subtracting n times Row 2 from Row 4. The new tableau is

x	y	r	s	u	t	Q	
1	−1	1	0	0	0	0	4
2	1	0	−1	1	0	0	4
4	−1	0	0	0	1	0	8
$1-2n$	$2-n$	0	n	0	0	1	$-4n$

We may now begin the simplex procedure. Our choice is to leave n as it is and use some algebra or substitute some large number for n and just use arithmetic. Since this technique is designed for use on computers which do not use algebra, we shall follow their lead. In this problem, $n = 20$ is considerably larger than any other number so let us make this substitution. The largest negative entry is in the x-column. The possible ratios are 4/1, 4/2, 8/4. Since the last two are equal we may pick either the 2 or the 4 as the pivot. Let us take the 2, which simplifies fractions. We have

x	y	r	s	u	t	Q		
1	−1	1	0	0	0	0	4	
(1)	1/2	0	−1/2	1/2	0	0	2	Row 2 ÷ 2
4	−1	0	0	0	1	0	8	
−39	−18	0	20	0	0	1	−80	

x	y	r	s	u	t	Q		
0	−3/2	1	1/2	−1/2	0	0	2	Row 1 − Row 2
1	1/2	0	−1/2	1/2	0	0	2	Row 2
0	−3	0	2	−2	1	0	0	Row 3 − 4 × Row 2
0	3/2	0	1/2	39/2	0	1	−2	Row 4 + 39 × Row 2

For each unknown, the last entry is now positive; the process stops and we read $x = 2$, $r = 2$, $t = 0$, $y = 0$, $s = 0$, $u = 0$, $Q = -2$. Since r, s, t are slack variables, the pertinent information is $x = 2$, $y = 0$ and $Q = P = -C = -2$, that is, $C = 2$.

EXERCISES

1. Redo the example in the text assuming that 4 were chosen as the pivot.
2. Use the simplex procedure to

$$\begin{aligned} \text{Minimize:}\quad & x + 4y \\ \text{Subject to:}\quad & 4x + y \geq 5 \\ & 5x - y \leq 4 \\ & -x + 2y \leq 10 \\ & x, y \geq 0 \end{aligned}$$

3. Use the simplex procedure to

$$\begin{aligned} \text{Minimize:}\quad & x + 7y \\ \text{Subject to:}\quad & 2x + y \geq 10 \\ & x + y \geq 9 \\ & 2x + 5y \geq 24 \\ & x, y \geq 0 \end{aligned}$$

4. Use the simplex procedure to

$$\begin{aligned} \text{Minimize:}\quad & 2x - 3y \\ \text{Subject to:}\quad & 2x - y - z \geq 3 \\ & x - y + z \geq 2 \\ & x, y, z \geq 0 \end{aligned}$$

5. Use the simplex procedure to

$$\begin{aligned} \text{Minimize:}\quad & -x + 2y \\ \text{Subject to:}\quad & 5x - 2y \leq 3 \\ & x + y \geq 1 \\ & -3x + y \leq 3 \\ & -3x - 3y \leq 2 \\ & x, y \geq 0 \end{aligned}$$

6

Recurrences and Progressions

6.1 The Geometric Progression

At Metropolitan College every student must pass basic mathematics. Fully one-half of the students in the course fail each term but no one gives up. Accordingly, we have a Markov process with two states, s_1 = taking basic math, s_2 = finished basic math. The transition matrix for the process is

$$\mathbf{P} = \begin{pmatrix} 1/2 & 1/2 \\ 0 & 1 \end{pmatrix}$$

Since everyone starts off taking basic math, the probability of being in State 1 initially is 1 and of being in State 2 initially is 0. Hence,

$$\mathbf{v}_0 = (1, 0)$$

Letting $\mathbf{v}_k = (p_k, q_k)$ be the probabilities of the two states after k semesters, we have

$$\mathbf{v}_{k+1} = \mathbf{v}_k\mathbf{P}$$

That is,

$$(p_{k+1}, q_{k+1}) = [(1/2)p_k, (1/2)p_k + q_k]$$

$$p_{k+1} = (1/2)p_k \quad \textbf{(1)}$$

$$q_{k+1} = q_k + (1/2)p_k \quad \textbf{(2)}$$

Equations (1) and (2) are called *recurrence relationships*. The first one is particularly simple since $p_0 = 1$. Letting $k = 0$, we have

$$p_1 = (1/2)p_0 = (1/2)(1) = (1/2)$$

and letting $k = 1$, we have

$$p_2 = (1/2)p_1 = (1/2)(1/2) = 1/4 = (1/2)^2$$

Letting $k = 2$ yields

$$p_3 = (1/2)p_2 = (1/2)(1/2)^2 = (1/2)^3$$

and, in general,

$$p_k = (1/2)^k$$

These numbers are said to form a geometric progression. The characterization of a geometric progression, $u_1, u_2, u_3, \ldots$, is that

$$u_{k+1} = ru_k$$

Here r is called the *common ratio* since $u_2/u_1 = r$; $u_3/u_2 = r$; $u_4/u_3 = r$; . . . . If $u_0 = a$, then $u_1 = ar$; $u_2 = ar^2$; $u_3 = ar^3$, ... In our case $a = 1$ and $r = 1/2$.

In planning schedules, we must know how many students are expected to be taking the course in any given term. Let us suppose that the school admits 1,000 students each term. Let S_k be the expected number of students taking the course in the kth term. Thus,

$$
\begin{aligned}
S_1 &= 1{,}000 \\
S_2 &= 1{,}000 + (1/2)S_1 = 1{,}000 + 500 \\
S_3 &= 1{,}000 + (1/2)S_2 = 1{,}000 + (1/2)(1{,}000 + 500) \\
&\qquad\qquad\qquad\quad = 1{,}000 + 500 + 250 \\
S_4 &= 1{,}000 + (1/2)S_3 = 1{,}000 + 500 + 250 + 125 \\
&\vdots \\
S_k &= a + ar + ar^2 + \cdots + ar^{k-1} \quad \text{where } a = 1{,}000,\ r = 1/2
\end{aligned}
$$

We see that S_k is the sum of the first k terms of the geometric progression. Notice that the first k terms start with a and end with ar^{k-1}. It would be nice to have a simple formula for this expression. To find it we write

$$
\begin{aligned}
S_k &= a + ar + ar^2 + \cdots + ar^{k-1} \\
rS_k &= \qquad ar + ar^2 + \cdots + ar^{k-1} + ar^k \\
S_k - rS_k &= a - ar^k \\
(1 - r)S_k &= a(1 - r^k) \\
S_k &= \frac{a(1 - r^k)}{1 - r}
\end{aligned}
$$

In our previous example, $a = 1{,}000$, $r = 1/2$,

$$
S_2 = \frac{1{,}000(1 - 1/4)}{1 - 1/2} = \frac{1{,}000(3/4)}{1/2} = 1{,}500
$$

It was easier to do that one directly without the formula. However, what is our prediction for the tenth term?

$$
\begin{aligned}
S_{10} &= \frac{1{,}000(1 - (1/2)^{10})}{1 - 1/2} = \frac{1{,}000(1 - (1/2)^{10})}{1/2} \\
&= 2{,}000[1 - (1/2)^{10}] = 2{,}000 - 2{,}000/2^{10}
\end{aligned}
$$

Since $2^{10} = 1{,}024$, S_{10} is about 1,998.

We may also use this formula to check the value of q_k in Formula (2). Recall that

$$
q_{k+1} = q_k + (1/2)p_k
$$

Since

$$
p_k = (1/2)^k
$$

and since $q_0 = 0$,

$$\begin{aligned} q_1 &= 0 + 1/2 = 1/2 \\ q_2 &= 1/2 + (1/2)^2 \\ q_3 &= (1/2) + (1/2)^2 + (1/2)^3 \\ &\vdots \\ q_k &= 1/2 + (1/2)^2 + (1/2)^3 + \cdots + (1/2)^k \end{aligned}$$

Thus q_k is the sum of the first k terms of a geometric progression with $a = 1/2$ and $r = 1/2$. Therefore

$$q_k = \frac{(1/2)(1 - (1/2)^k)}{1 - 1/2} = 1 - (1/2)^k$$

Of course, we knew this previously since

$$\mathbf{v}_k = (p_k, q_k)$$

which means $p_k + q_k = 1$ and we already had $p_k = (1/2)^k$.

What is interesting is that as k becomes large, $p_k = (1/2)^k$ gets very small. Thus when $k = 10$, $p_{10} = (1/2)^{10}$, which is about 0.001. Similarly p_{20} is about 0.000001. We describe this behavior by saying that p_k approaches zero as k approaches infinity (becomes very large).*

For $q_k = 1 - (1/2)^k$, we say q_k approaches one as k approaches infinity. In the previous example $S_k = 2{,}000 - 2{,}000/2^k$ approaches 2,000 as k approaches infinity. In general, if r is between -1 and $+1$, then for large values of k, r^k will approach zero. Hence, we have

$$S_k = \frac{a(1 - r^k)}{1 - r}$$

approaches

$$S = \frac{a}{1 - r}$$

as k approaches infinity. We shall call

$$S = \frac{a}{1 - r}$$

---

* The author apologizes for using the term *infinity* in a book on finite mathematics, but it seems unavoidable in this section.

the sum of the *infinite geometric progression.* This is the value which S_k approaches for large values of k, but remember that the formula is nonsense unless $-1 < r < 1$. Some examples follow.

Example 1 3, -12, 48 are the first three terms of a geometric progression. Find r, find the sixth term and find the sum of the first six terms.

We have $a = 3$, $ar = -12$; thus $r = -4$. The sixth term is $ar^5 = 3(-4)^5 = -3{,}072$. The sum of the first six terms is

$$S_6 = \frac{a(1 - r^6)}{1 - r} = \frac{3(1 - (-4)^6)}{1 - (-4)} = -2{,}457$$

If we computed $S = a/(1 - r) = 3/5$ as the sum of the infinite progression, we see that the result is nonsense.

Example 2 In a geometric progression, the first term is 3 and the fourth term is 8/9. Find the second and third terms and the sum of the infinite progression.

We have $a = 3$, $ar^3 = 8/9$. Thus,

$$r^3 = 8/27 \qquad r = 2/3$$

Hence the progression goes

$$3, 2, 4/3, 8/9, \ldots$$

and

$$S = \frac{3}{1 - 2/3} = 9$$

EXERCISES

1. Referring to the basic math course at Metropolitan College as described in the text, suppose that the school only admits new students in the fall semester. If the entering class is 1,000 students, find the numbers expected to be in basic math in the fall semester and spring semester for the first three years. Find the long-run expected numbers.
2. Referring to the basic math course at Metropolitan College as described in the text, suppose the fail rate is reduced to 1/3. Find $\mathbf{v}_k$ and find the long-run expected number in the course if 1,000 students are admitted to the college each semester.

3. Repeat Exercise 1, if the fail rate is 1/3.
4. The numbers 4, x, 16 are in geometric progression. Find the possible values of x.
5. The numbers x, 8, x^2 are in geometric progression. Find the possible values of x.
6. The first term of a geometric progression is 162 and the fifth is 32. Find the possible values of r and the corresponding sums of infinite progression.
7. The first term of an infinite geometric progression is 6 and the sum is 7. Find r and the first four terms.
8. The first five terms of a geometric progression sum to 5. The infinite sum is 7; find the first four terms.
9. The numbers 4, x, y, z, 324 form a geometric progression. Find x, y, z.
10. Given the recurrence $u_{k+1} = 2u_k + 1$, $u_0 = 1$, find the $u_1, u_2, \ldots, u_5$. Can you guess a general formula for u_k?
11. Repeat Exercise 10 for $u_{k+1} = 2(u_k - 1)$, $u_0 = 3$.
12. Repeat Exercise 10 for $u_{k+2} = u_{k+1} - u_k$, $u_0 = 1$, $u_1 = 2$.

6.2 Application to Markov Chains

Let us look at our basic math Markov process again. We have

$$\mathbf{v}_k = [(1/2)^k, 1 - (1/2)^k] = (p_k, q_k)$$

As k becomes large, p_k approaches zero and q_k approaches one. Since q_k is the probability that a student has completed basic math after k terms, this says that the probability of eventually completing the course approaches one. Or in other words, if the student remains long enough, he is virtually certain to pass. Of course, an interesting question is, How long should one expect it to take? That is, how many terms will it take the average student to complete the course? If we denote this number by N, then $N = E(X)$ where X is the number of trials needed to pass the course. Now, we have never found the expected value of a random variable which can take on an infinite number of possible values. However, it turns out that the same formula will work, that is,

$$E(X) = 1p_1 + 2p_2 + 3p_3 + \cdots$$

Believe it or not, this can be evaluated. To do so, we note that if $r = 1/2$, the sum of the infinite geometric progression is

$$S = \frac{a}{1 - r} = 2a$$

Now,

$$\begin{aligned}
E(X) &= p_1 + 2p_2 + 3p_3 + 4p_4 + \cdots \\
E(X) &= (1/2) + 2(1/2)^2 + 3(1/2)^3 + 4(1/2)^4 + \cdots \\
E(X) &= (1/2) + (1/2)^2 + (1/2)^3 + (1/2)^4 + \cdots \\
&\qquad\quad + (1/2)^2 + (1/2)^3 + (1/2)^4 + \cdots \\
&\qquad\qquad\qquad + (1/2)^3 + (1/2)^4 + \cdots \\
&\qquad\qquad\qquad\qquad\quad + (1/2)^4 + \cdots \\
&\qquad\qquad\qquad\qquad\qquad\qquad \vdots
\end{aligned}$$

In the last equation, each line is a geometric progression with $r = 1/2$. In the first line $a = 1/2$, in the second $a = (1/2)^2$, and so on. Hence

$$E(X) = 2(1/2) + 2(1/2)^2 + 2(1/2)^3 + \cdots$$

which is a geometric progression with $r = 1/2$ and $a = 1$. Thus, $E(X) = 2(1) = 2$. That is, the average student will spend two terms taking basic math. Intuitively, this seems correct. If there is a 50% chance of passing each time he tries, then the average person should pass in two tries.

Actually we can attack this problem without resorting to the infinite progression. Let N be the number of trials a student is expected to take. Since everyone starts in State 1, we can say that N is equal to 1 plus the average number of additional trials needed. Now 1/2 the students pass and need no more tries. The remaining 1/2 start over and need an average of N more trials. Hence,

$$\begin{aligned}
N &= 1 + (1/2)(0) + (1/2)N \\
N &= 1 + (1/2)N \\
(1/2)N &= 1 \\
N &= 2
\end{aligned}$$

Please notice that in this hypothetical example, the student who is repeating the course has no more chance of passing than the newcomer.

This type of Markov chain is characterized by the existence of an *absorbing state*, that is, a state which once entered is never left. In this case, s_2 is an absorbing state. It is easily recognized by the fact that the entry in Row 2, Column 2 of the transition matrix is a 1. Since there is a nonzero probability that the process enters s_2 and there is no probability of ever leaving it, eventually the process must end in State 2.

Now suppose that in order to reduce the number of failures in basic math, Metropolitan College divides the one-semester course into two one-semester courses, allowing more time for study. Thus, we have Math I and Math II. Further, suppose that in each course 1/3 of the students fail and have to repeat. Using the obvious notation, we have as the transition matrix for this process

$$\mathbf{P} = \begin{pmatrix} 1/3 & 2/3 & 0 \\ 0 & 1/3 & 2/3 \\ 0 & 0 & 1 \end{pmatrix}$$

Furthermore, suppose that an exemption examination is given to entering students, on the basis of which a student may be placed in Math I or in Math II or he may be excused from math. This gives as an initial probability vector

$$\mathbf{v}_0 = (p_1, p_2, p_3)$$

which shows the likelihoods that an entering student selected at random is in s_1, s_2 or s_3, respectively. This new situation raises a series of questions. For example,

1. Since we now have two nonabsorbing states, is there still probability 1 that the process will be absorbed in s_3?
2. Assuming that it must be absorbed in s_3, what is the expected number of steps until absorption?
3. What is the expected number of times that the process will be in s_1 and s_2 given that it starts in s_1? Given that it starts in s_2?

To answer these questions, consider the matrix $\mathbf{P}$. The four elements in the upper lefthand section form a matrix:

$$\mathbf{Q} = \begin{pmatrix} 1/3 & 2/3 \\ 0 & 1/3 \end{pmatrix}$$

which displays the transition probabilities between the nonabsorbing states s_1 and s_2. The largest entry in $\mathbf{Q}$ is 2/3. Therefore, after one step the probability that the process will be in s_1 or s_2 is at most 2/3. After two steps, this probability will be at most $(2/3)^2$; after three steps $(2/3)^3$ and so on.

Since the nth power of 2/3 rapidly tends to zero, the process must eventually be absorbed in s_3.

Now let n_{ij} be the expected number of steps during which the process will be in state s_j given that it starts in state s_i, that is,

$$n_{11} = \text{time spent in } s_1, \text{ given we start in } s_1$$
$$n_{12} = \text{time spent in } s_2, \text{ given we start in } s_1$$
$$n_{21} = \text{time spent in } s_1, \text{ given we start in } s_2$$
$$n_{22} = \text{time spent in } s_2, \text{ given we start in } s_2$$

Let

$$\mathbf{N} = \begin{pmatrix} n_{11} & n_{12} \\ n_{21} & n_{22} \end{pmatrix}$$

We may now write

$$\mathbf{N} = \mathbf{I} + \mathbf{QN}$$

analogously to the previous example. Why should this be so? Suppose $\mathbf{u}_0 = (p_1, p_2)$ gives the initial probabilities of s_1 and s_2. Then $\mathbf{u}_0\mathbf{N}$ gives the expected number of times spent in s_1 and s_2. However, suppose one step has passed. Now the probabilities of the two states is given by $\mathbf{u}_0\mathbf{Q}$ and the expected number of remaining times in the two states becomes $(\mathbf{u}_0\mathbf{Q})\mathbf{N}$. Logically, the total number of times spent in a state = number of times spent in that state on the first step + number of times spent in that state on the remaining trials. Now, the expected number of times spent in a given state on the first step is simply equal to the probability of being in that state on the first step. Hence

$$\mathbf{u}_0\mathbf{N} = \mathbf{u}_0 + (\mathbf{u}_0\mathbf{Q})\mathbf{N}$$

That is,

$$\mathbf{u}_0\mathbf{N} = \mathbf{u}_0(\mathbf{I} + \mathbf{QN})$$

But this is true for every possible $\mathbf{u}_0$ and hence

$$\mathbf{N} = \mathbf{I} + \mathbf{QN}$$

Solving for $\mathbf{N}$, we obtain

$$\mathbf{N} - \mathbf{QN} = \mathbf{I}$$
$$(\mathbf{I} - \mathbf{Q})\mathbf{N} = \mathbf{I}$$
$$\mathbf{N} = (\mathbf{I} - \mathbf{Q})^{-1}$$

assuming that $(\mathbf{I} - \mathbf{Q})^{-1}$ exists (which it always will for an absorbing chain). For the case in question,

$$\mathbf{I} - \mathbf{Q} = \begin{pmatrix} 1 & 0 \\ 0 & 1 \end{pmatrix} - \begin{pmatrix} 1/3 & 2/3 \\ 0 & 1/3 \end{pmatrix} = \begin{pmatrix} 2/3 & -2/3 \\ 0 & 2/3 \end{pmatrix}$$

$$= (2/3)\begin{pmatrix} 1 & -1 \\ 0 & 1 \end{pmatrix}$$

and

$$\mathbf{N} = (\mathbf{I} - \mathbf{Q})^{-1} = (3/2)\begin{pmatrix} 1 & 1 \\ 0 & 1 \end{pmatrix} = \begin{pmatrix} 3/2 & 3/2 \\ 0 & 3/2 \end{pmatrix}$$

Thus, given that the process starts in s_1, we may expect that it will stay in s_1 for 1.5 steps and in s_2 for 1.5 steps. If it starts in s_2 we expect it to spend zero steps in s_1 and 1.5 steps in s_2. The zero steps in s_1 must be correct since there is no way to go from s_2 into s_1 in this example. Clearly the expected number of steps until absorption is three if we start in s_1 and 1.5 if we start in s_2. That is, the expected numbers of steps until absorption are the sums of the rows of **N**. We know from past examples that this is simply **NU** where **U** is a column vector of all ones. Thus,

$$\mathbf{t} = \mathbf{NU}$$

gives the expected numbers of steps until absorption starting from each of the nonabsorbing states.

An interesting sidelight is the question, Does this new math-sequence really help the student to finish earlier? Not the student who starts at the beginning. Although his probability of failure has been reduced from 1/2 to 1/3, he now expects to need three terms to finish the sequence, whereas he expected before to need only two terms. However, on the average the student may be better off if enough are able to start in s_2 or s_3 after the placement examination.

Let us further complicate the model under consideration. Suppose it is found that 10% of all students drop out of school each semester while taking Math I and II, never to return. We let

s_1 = taking Math I

s_2 = taking Math II

s_3 = completed mathematics

s_4 = dropped out before completing mathematics

Statistics show the transition matrix to be

$$\mathbf{P} = \begin{pmatrix} .3 & .6 & 0 & .1 \\ 0 & .3 & .6 & .1 \\ 0 & 0 & 1 & 0 \\ 0 & 0 & 0 & 1 \end{pmatrix}$$

Now, we may visualize **P** as

$$\mathbf{P} = \begin{pmatrix} \mathbf{Q} & \mathbf{R} \\ \mathbf{Z} & \mathbf{I} \end{pmatrix}$$

where **Q** gives transition probabilities between nonabsorbing states and **R** gives transition probabilities from nonabsorbing to absorbing states. A reasonable question to ask at this point is, What are the probabilities of eventual absorption in each of the absorbing states?

Let **B** be a matrix giving the probabilities of absorption in s_3 and s_4 given that the process starts in s_1 and s_2, that is,

$$\mathbf{B} = \begin{pmatrix} b_{13} & b_{14} \\ b_{23} & b_{24} \end{pmatrix}$$

where b_{ij} is the probability of absorption in State j having started in State i. Now the probability of eventual absorption must be equal to the probability of immediate absorption plus the probability of transition to another nonabsorbing state, followed by eventual absorption. Thus, for example,

$$b_{14} = 0.1 + (0.3)b_{14} + (0.6)b_{24}$$

Immediate absorption	Remain in s_1 followed by eventual absorption	Transition to s_2 followed by eventual absorption

Written in matrix form, this relationship becomes

$$\begin{aligned} \mathbf{B} &= \mathbf{R} + \mathbf{QB} \\ \mathbf{B} - \mathbf{QB} &= \mathbf{R} \\ (\mathbf{I} - \mathbf{Q})\mathbf{B} &= \mathbf{R} \\ \mathbf{B} &= (\mathbf{I} - \mathbf{Q})^{-1}\mathbf{R} \end{aligned}$$

Since $(\mathbf{I} - \mathbf{Q})^{-1} = \mathbf{N}$, we have finally

$$\mathbf{B} = \mathbf{NR}$$

Summarizing, if we have an absorbing Markov chain, we first will write its transition matrix in the form

$$\mathbf{P} = \begin{pmatrix} \mathbf{Q} & \mathbf{R} \\ \mathbf{Z} & \mathbf{I} \end{pmatrix}$$

To do this, we may have to renumber the states so that all the absorbing states are last. We next find the "fundamental matrix," $\mathbf{N} = (\mathbf{I} - \mathbf{Q})^{-1}$. Then $\mathbf{N}$ gives the expected times in each nonabsorbing state, $\mathbf{t} = \mathbf{NU}$ gives the expected times until absorption and $\mathbf{B} = \mathbf{NR}$ gives the probabilities of absorption in the several absorbing states. Thus, taking our current example, we have

$$\mathbf{P} = \begin{pmatrix} .3 & .6 & 0 & .1 \\ 0 & .3 & .6 & .1 \\ 0 & 0 & 1 & 0 \\ 0 & 0 & 0 & 1 \end{pmatrix}$$

$$\mathbf{Q} = \begin{pmatrix} .3 & .6 \\ 0 & .3 \end{pmatrix} \qquad \mathbf{I} - \mathbf{Q} = \begin{pmatrix} .7 & -.6 \\ 0 & .7 \end{pmatrix}$$

$$\mathbf{N} = (\mathbf{I} - \mathbf{Q})^{-1} = (10/7)\begin{pmatrix} 1 & 6/7 \\ 0 & 1 \end{pmatrix}$$

$$\mathbf{B} = \mathbf{NR} = (10/7)\begin{pmatrix} 1 & 6/7 \\ 0 & 1 \end{pmatrix}\begin{pmatrix} 0 & 1/10 \\ 6/10 & 1/10 \end{pmatrix}$$

$$\mathbf{B} = \begin{pmatrix} 36/49 & 13/49 \\ 6/7 & 1/7 \end{pmatrix}$$

Analyzing these matrices, we draw the following conclusions: Reading the first line of $\mathbf{B}$, we see that of students starting in Math I, 36/49, or about 73%, will be absorbed in the first absorbing state. That is, they will finish the sequence. The remaining 13/49, about 27%, will drop out before finishing. For students who begin in Math II, these probabilities are 6/7 (86%) and 1/7 (14%), respectively. From the matrix $\mathbf{N}$, we get the following information: Students who start in Math I may expect to spend 10/7 or 1.43 semesters in Math I and 60/49 or 1.22 semesters in Math II on the average. Thus, students starting in Math I may expect to average 2.65 semesters in mathematics courses before either finishing up or dropping out of school. For students starting in Math II, we see no terms spent in

Math I and 10/7 terms spent in Math II. Of course, among nondropouts the average time until absorption is somewhat longer. Remember, we take into account no relationship between course failure and dropout likelihood. A more realistic model would include such input. In fact, a more realistic model would almost surely not be a Markov process. Nevertheless, a Markov process does give a simple and fairly accurate approximation.

In the analysis of absorbing chains, we have ignored the role of the initial probabilities of the various states. That is, **N**, **t** and **B** do not take the initial probabilities into account. To do so, recall that the ith row of **N** gives the expected times spent in each nonabsorbing state, assuming that the process starts in each nonabsorbing state. Again, n_{ij} is the expected time spent in s_j given that the process starts in s_i. Hence, if $\mathbf{v}_0 = (p_1, p_2, \ldots, p_k)$ gives the initial probabilities of the nonabsorbing states, then $\mathbf{v}_0\mathbf{N}$ gives the average time spent in each nonabsorbing state. Similarly, $\mathbf{v}_0\mathbf{t}$ yields the average time until absorption, and $\mathbf{v}_0\mathbf{B}$ gives the probability of ending in each of the absorbing states.

EXERCISES

1. Consider the four-state model in the text. How long does the average student stay in math if
   (a) all students start in Math I?
   (b) half the students start in Math I and half in Math II?
   (c) 4/5 start in Math I and 1/5 in Math II?
   (d) 1/3 start in each Math I and Math II and 1/3 exempt math?

2. Consider the four-state model in the text. What percent of the students finish math and what percent drop out under each of initial conditions described in Exercise 1?

3. Given

$$\mathbf{P} = \begin{pmatrix} .8 & .1 & 0 & .1 & 0 \\ .4 & 0 & .4 & .1 & .1 \\ .6 & .1 & .1 & .1 & .1 \\ 0 & 0 & 0 & 1 & 0 \\ 0 & 0 & 0 & 0 & 1 \end{pmatrix}$$

   find
   (a) the expected times until absorption starting from s_1, s_2 and s_3;
   (b) the probability of absorption in s_4 and s_5 starting from each nonabsorbing state.

4. A particle moves along the x-axis moving 1 unit to the left with probability $1/2$ or 1 unit to the right with probability $1/2$ on each step. The particle may start at Point 2 or 3. If it reaches 1 or 4 it gets absorbed.
   (a) Write a transition matrix for the process.
   (b) Put the matrix in standard form.
   (c) Analyze the process.
   (This is called a *random walk with absorbing boundaries.*)

5. Repeat Exercise 4 for the case where absorption is at 1 and 5 and the process may start at 2, 3 or 4.

6. Figure 1 shows a maze in which a mouse is placed. The mouse may be placed in Compartment 1, 2 or 3. He selects doors randomly, changing locations every second. If he goes through any exit, the process ends. Set up the transition matrix for this as a Markov process and analyze in terms of number of seconds expected to take until escape and likelihood of escape through each exit.

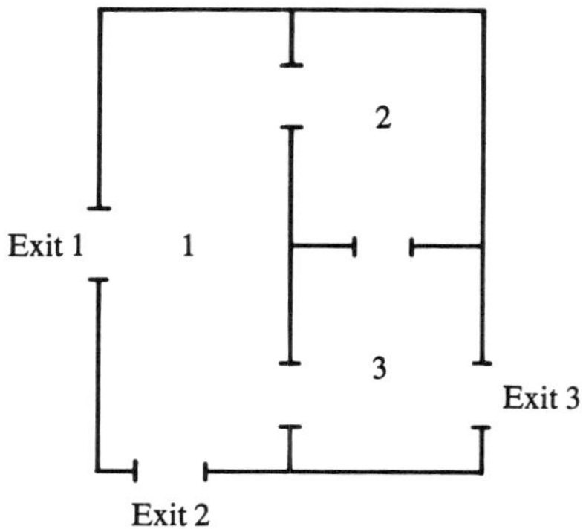

Figure 1

7. Repeat Exercise 6 for the mazes in figure 2.

8. Analyze the Markov chains having the following transition matrices:

(a) $\begin{pmatrix} 1/4 & 3/4 \\ 0 & 1 \end{pmatrix}$

(b) $\begin{pmatrix} 1/2 & 1/4 & 1/4 \\ 1/4 & 1/4 & 1/2 \\ 0 & 0 & 1 \end{pmatrix}$

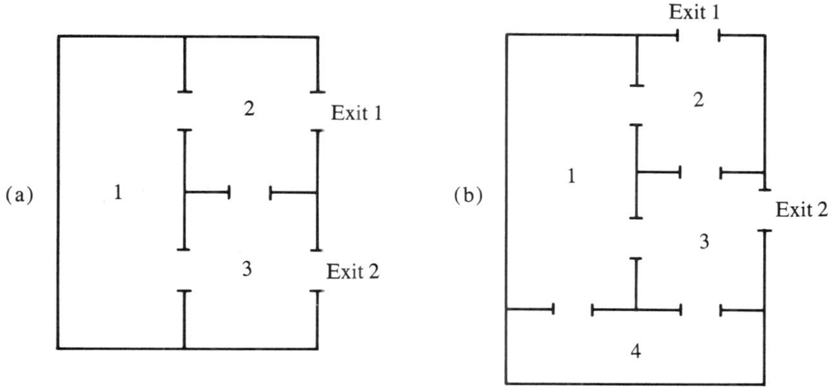

Figure 2

(c) $\begin{pmatrix} 1/4 & 1/4 & 1/4 & 1/4 \\ 1/4 & 1/2 & 0 & 1/4 \\ 0 & 0 & 1 & 0 \\ 0 & 0 & 0 & 1 \end{pmatrix}$

6.3 Arithmetic Progressions

What property is shared by the following sequences:

$$1, 3, 5, 7, 9, 11, \ldots$$

$$-1, 2, 5, 8, 11, 14, \ldots$$

$$1, 1\tfrac{1}{2}, 2, 2\tfrac{1}{2}, 3, 3\tfrac{1}{2}, \ldots$$

$$100, 75, 50, 25, 0, -25, \ldots$$

In each case, the difference between successive terms is a constant, that is, $u_{n+1} - u_n = d$ where d is called the *common difference*. In the first case, $d = 2$; in the second, $d = 3$; in the third, $d = 1/2$; and in the fourth, $d = -25$. This type of sequence is called an *arithmetic progression*. It is easy to work out the general properties of such a progression. The fundamental formula may be rewritten,

$$u_{n+1} = u_n + d$$

If we call the first term a, then

$$\begin{aligned}
u_1 &= a \\
u_2 &= a + d \\
u_3 &= a + d + d = a + 2d \\
u_4 &= a + 2d + d = a + 3d \\
&\vdots \\
u_n &= a + (n - 1)d
\end{aligned}$$

In the sequence $-1, 2, 5, 8, \ldots$, we have $a = -1$, $d = 3$ and the nth term is $u_n = -1 + 3(n - 1)$.

To find the sum of an arithmetic progression, write

$$\begin{aligned}
S_n &= a + (a + d) + (a + 2d) + \cdots + [a + (n - 1)d] \\
S_n &= na + d + 2d + 3d + \cdots + (n - 1)d \\
S_n &= na + [1 + 2 + 3 + \cdots + (n - 1)]d
\end{aligned}$$

We now must sum $X_n = 1 + 2 + 3 + \cdots + (n - 2) + (n - 1)$. To do so, we shall write X_n twice, first with the terms in increasing order and second with the terms in decreasing order. Thus

$$\begin{aligned}
X_n &= 1 + 2 + 3 + \cdots + (n - 2) + (n - 1) \\
X_n &= (n - 1) + (n - 2) + (n - 3) + \cdots + 2 + 1
\end{aligned}$$

Add the two equations:

$$2X_n = \underbrace{n + n + \cdots + n}_{(n-1)\text{ times}} = n(n - 1)$$

Thus,

$$\begin{aligned}
X_n &= \frac{n(n - 1)}{2} \\
S_n &= na + \frac{n(n - 1)\,d}{2} \\
&= n\left[a + \frac{(n - 1)d}{2}\right] \\
&= n\left[\frac{2a + (n - 1)d}{2}\right] = n\left[\frac{a + u_n}{2}\right]
\end{aligned}$$

That is, the sum of the first n terms is the average of the first and last terms multiplied by the number of terms.

Suppose we take the progression 100, 75, 50, 25. . . .; the tenth term is $u_{10} = 100 + 9(-25) = -125$ and the sum of first ten terms,

$$S_{10} = \frac{10(100 - 125)}{2} = -125$$

EXERCISES

1. Find the fifteenth term of sequence 2, 5, 8, 11, 14, ... Find the sum of the first 45 terms.
2. The third term of an arithmetic progression is 3 and the fifth term is -2. Find the fourth term.
3. The nth term of an arithmetic progression is a and the $(n + 2)$nd term is b. Find the $(n + 1)$st term.
4. The nth term of a progression is $u_n = 2 + 5(n - 1) + 3^n$. Find the sum of the first ten terms.
5. The fourth term of an arithmetic progression is -1 and the ninth term is 14. Find the first term and the sum of the first thirty terms.
6. The fourth term of an arithmetic progression is 13 and the seventh term is 22. Find the first term, the common difference and the sum of the first sixty terms.
7. Find the sum of all the even numbers from 22 to 200, both inclusive.
8. The first term of an arithmetic progression is 6 and the sum of the first twelve terms is 20. Find the common difference.
9. An experiment has ten possible outcomes, the probabilities of which form an arithmetic progression. If the least likely outcome has probability 0.02, what is the probability of the most likely? Find the sum of the probabilities of the five least likely outcomes.
10. The sum of the first seven terms of an arithmetic progression is 15 and the sum of the first eleven terms is 10. Find the first four terms of the progression.

6.4 Other Recurrences

We have considered two common types of recurrence relationships:

$$u_{n+1} - u_n = d$$
$$au_{n+1} - bu_n = 0$$

The first of these gives rise to the arithmetic progression, the second to the geometric. Let us consider some more general relationships, such as

$$u_{n+1} - u_n = a_n$$

where a_n is some known function of n. For example,

$$u_{n+1} - u_n = n$$

Writing this as $u_{n+1} = u_n + n$, and letting $u_0 = h$, we obtain

$$\begin{aligned} u_1 &= 1 + h \\ u_2 &= 2 + 1 + h = 3 + h \\ u_3 &= 3 + 3 + h = 6 + h \\ u_4 &= 4 + 6 + h = 10 + h \\ &\vdots \end{aligned}$$

In general, $u_n = h + (1 + 2 + 3 + \cdots + n)$. We have already seen that $1 + 2 + 3 + \cdots + (n - 1) = n(n - 1)/2$. Therefore, $1 + 2 + 3 + \cdots + n = n(n + 1)/2$ (obtained by adding n to both sides or by simply substituting $n + 1$ for n in the equation). More generally, we can see that if

$$u_n = an^2 + bn + c$$

then

$$u_{n+1} - u_n = 2an + a + b$$

That is, if u_n is quadratic in n, then $u_{n+1} - u_n$ is linear in n. Hence, if we wish to solve

$$u_{n+1} - u_n = dn + e$$

we should try to find a solution of the form $u_n = an^2 + bn + c$. For example, let us solve

$$u_{n+1} - u_n = 3n + 2$$

Let $u_n = an^2 + bn + c$, substituting

$$\begin{aligned} [a(n + 1)^2 + b(n + 1) + c] - [an^2 + bn + c] &= 3n + 2 \\ an^2 + 2an + a + bn + b + c - an^2 - bn - c &= 3n + 2 \\ 2an + a + b &= 3n + 2 \end{aligned}$$

Since this must be true for all n, it must be true for $n = 0$; hence $a + b = 2$. It must also be true for $n = 1$; hence $3a + b = 5$. Solving these two simultaneously, we obtain $a = 3/2$, $b = 1/2$. Thus, the solution to

$$u_{n+1} - u_n = 3n + 2$$

is

$$u_n = (3/2)n^2 + (1/2)n + c$$

for any number c. If, in addition, we had $u_0 = 1$, then

$$1 = 0 + 0 + c \qquad c = 1$$

and

$$u_n = (3/2)n^2 + (1/2)n + 1$$
$$u_0 = 1;\ u_1 = 3;\ u_2 = 8;\ldots$$

By the way, it is not really necessary to solve simultaneous equations to find a and b. In general, if two polynomials in n agree for all n, then the coefficients of the powers must be equal. So from $2an + (a + b) = 3n + 2$ we may read immediately

$$2a = 3 \qquad \text{(Coefficients of } n \text{ are identical.)}$$
$$a + b = 2 \qquad \text{(Constants are identical.)}$$

The pattern should be obvious; if we want to solve

$$u_{n+1} - u_n = en^2 + fn + g$$

we should try

$$u_n = an^3 + bn^2 + cn + d$$

and so on for higher-degree polynomials. One other common form for a_n is r^n. Thus,

$$u_{n+1} - u_n = 3^n$$

If we try $u_n = a(3^n)$,

$$a(3^{n+1}) - a(3^n) = 3^n$$
$$3a(3^n) - a(3^n) = 3^n$$
$$3a - a = 1$$
$$a = 1/2$$

Hence,

$$u_n = (1/2)(3^n);\ u_0 = (1/2),\ u_1 = (1/2)(3),\ u_2 = (1/2)(9), \ldots$$

is a solution.

However, suppose we required $u_0 = 4$. Since $u_{n+1} = u_n + 3^n$, the solution must be

$$\begin{aligned} u_0 &= 4 \\ u_1 &= 4 + 3^0 = 5 \\ u_2 &= 5 + 3^1 = 8 \\ u_3 &= 8 + 3^2 = 17 \\ &\vdots \end{aligned}$$

Why have we not found this solution? The answer is that $u_n = (1/2)(3^n)$ is not the most general solution. In fact, if we take $u_n = (1/2)(3^n) + b$, this will also satisfy the equation. Now, we may satisfy the condition $u_0 = 4$ by letting $b = 3\frac{1}{2}$ and we have

$$u_n = (1/2)(3^n) + 3\tfrac{1}{2}$$

The trick is that any function which makes the left side of the equation zero may be added to $(1/2)(3^n)$ and the result will be a solution. In this case, the left side is $u_{n+1} - u_n$ which will only be zero for $u_n = b$ (constant). So try

$$u_{n+1} - 3u_n = 2 \qquad u_0 = 4$$

We will solve the problem in four steps:

1. We shall find $w_n = an + b$ which satisfies the equation;
2. We shall find a solution to $v_{n+1} - 3v_n = 0$;
3. We shall combine the solutions;
4. We shall pick constants to satisfy the condition $u_0 = 4$.

Starting with Step 1, let $w_n = an + b$ and substitute in the equation

$$\begin{aligned} [a(n + 1) + b] - 3[an + b] &= 2 \\ an + a + b - 3an - 3b &= 2 \\ -2an + a - 2b &= 2 \end{aligned}$$

Therefore,

$$\begin{aligned} -2a &= 0 \qquad a = 0 \\ a - 2b &= 2 \qquad b = -1 \end{aligned}$$

Hence, $w_n = -1$ is a solution. For Step 2, we see

$$v_{n+1} = 3v_n$$

which we recognize gives rise to the geometric progression

$$v_n = c(3^{n-1})$$

where c is arbitrary. For Step 3, combine the two answers to give

$$u_n = -1 + c(3^{n-1})$$

Step 4, $u_0 = 4$, yields

$$\begin{aligned} 4 &= -1 + c(3^{-1}) \\ 5 &= c/3 \\ c &= 15 \end{aligned}$$

This gives the final answer

$$u_n = -1 + 15(3^{n-1})$$

The reader may check by substituting in the original equation.

EXERCISES

1. Let $u_n = 1 + 2 + 3 + \cdots + n$. We may express this by the recurrence $u_{n+1} - u_n = n + 1$. Use the methods of this section to derive a formula for $1 + 2 + 3 + \cdots + n$.

2. Use the technique of Exercise 1 to find a formula for $1^2 + 2^2 + 3^2 + \cdots + n^2$.

3. Solve the recurrences
   (a) $u_{n+1} - u_n = 2 + 4n$; $u_0 = 2$;
   (b) $u_{n+1} - 2u_n = -2 + 4n$; $u_1 = 3$.

4. Solve the recurrences
   (a) $u_{n+1} - u_n = 2n + n^2$; $u_3 = 6$;
   (b) $u_{n+1} + 2u_n = 2n + n^2$; $u_0 = 2$.

5. Solve the recurrences
   (a) $u_{n+1} - u_n = 5^n$; $u_0 = 1$;
   (b) $u_{n+1} - 4u_n = (1/2)^n$; $u_1 = 1$.

6. Solve the recurrences
   (a) $2u_{n+1} - 3u_n = 5^n + 3;\ u_0 = 3;$
   (b) $u_{n+1} - 2u_n = 3^n - 2 - 2n;\ u_0 = 4;$
   (c) $u_{n+1} + 2u_n = 5(3^n) + 12(4^n) - 6;\ u_1 = 5.$

7. A bank deposit earns 5% interest every year. At the beginning of each year, I will deposit \$100 in the bank. The total amount in the bank after the nth year is u_n. Now,

$$u_{n+1} = (1.05)u_n + 100 \qquad u_0 = 0$$

   Find the general formula for u_n and find the total amount in the bank after ten years. (This is how one may calculate the accumulation of an annuity.)

8. One-third of all students taking Math I fail and retake the course. In addition, one thousand students from the entering class enroll. Let u_n be the number of students in the course in the nth semester. Use a recurrence to express u_{n+1} and solve for u_n.

9. One-quarter of all the students taking Math II fail and retake the course. In addition, the two-thirds of the students from Math I who passed go on to Math II and 500 students enter Math II via placement exam each term. Let v_n be the number of students in Math II in the nth semester. Express v_{n+1} in terms of v_n and u_n from Exercise 8. Use the result of Exercise 8 and solve for v_n.

6.5 Second-order Recurrences

Consider the following sequence:

$$4, 5, 9, 16, 26, 39, 55, 74, 96, \ldots$$

This sequence is neither arithmetic nor geometric. The differences between successive terms form the sequence

$$1, 4, 7, 10, 13, 16, 19, 22, \ldots$$

which is an arithmetic progression. Calling the first terms $u_1, u_2, \ldots$ and the terms $u_{n+1} - u_n = v_n$, we have

$$v_{n+1} - v_n = 3$$

which means

$$(u_{n+2} - u_{n+1}) - (u_{n+1} - u_n) = 3$$
$$u_{n+2} - 2u_{n+1} + u_n = 3$$

Previously, we have worked only problems involving u_{n+1} and u_n. Here we encounter a problem involving u_{n+2}, u_{n+1} and u_n. This is called a *second-order recurrence*. The earlier geometric and arithmetic examples were first-order recurrences.

To attack this problem, we shall extend the method of the preceding section. Since $v_{n+1} - v_n = 3$, v_n must be a linear expression in n, that is, $v_n = dn + e$. But $v_n = u_{n+1} - u_n$ and hence u_n of the form $an^2 + bn + c$ should provide a solution. In fact, let us try it.

$$[a(n + 2)^2 + b(n + 2) + c] - 2[a(n + 1)^2 + b(n + 1) + c] + [an^2 + bn + c] = 3$$

After some algebra, we obtain

$$2a = 3 \qquad a = 3/2$$

with b and c arbitrary. For first-order cases there was always one arbitrary constant which was determined by the value of u_0 (or some other given u_k). For the second-order case we have two arbitrary constants. Hence

$$u_n = (3/2)n^2 + bn + c$$

Since $u_1 = 4$ and $u_2 = 5$, we have

$$4 = 3/2 + b + c \qquad b + c = 5/2$$
$$5 = 6 + 2b + c \qquad 2b + c = -1$$

Therefore

$$b = -(7/2) \qquad c = 6$$

and

$$u_n = (3/2)n^2 - (7/2)n + 6$$

For more general second-order recurrences, such as

$$u_{n+2} - 5u_{n+1} + 6u_n = n + 2$$

we know that the first step is to look for a cubic expression that will satisfy the equation. Although algebraically difficult, the idea is simple. The harder problem is to find the function v_n such that

$$v_{n+2} - 5v_{n+1} + 6v_n = 0$$

The sought function must be added to the preceding cubic. The difficulty is that for the first-order case, we were able to recognize the geometric recurrence. In this case, we have no past experience to guide us.

However, reasoning by analogy, let us try the solution $v_n = cr^n$. We get

$$cr^{n+2} - 5cr^{n+1} + 6cr^n = 0$$

We could let $c = 0$ or $r = 0$, but that would give us only a trivial solution. Instead divide by cr^n, which yields

$$\begin{aligned} r^2 - 5r + 6 &= 0 \\ (r - 3)(r - 2) &= 0 \\ r = 3,\ r &= 2 \end{aligned}$$

Thus $c_1(3^n)$ and $c_2(2^n)$ are both solutions and the overall solution is

$$u_n = c_1(3^n) + c_2(2^n) + P_3$$

where P_3 is the cubic polynomial which we said we could find. Notice that the solution does involve two arbitrary constants as earlier mentioned. Actually, substituting $P_3 = an^3 + bn^2 + cn + d$, we obtain $a = b = 0$ and $P_3 = (1/2)n + 7/4$, giving a solution

$$u_n = c_1(3^n) + c_2(2^n) + (1/2)n + 7/4$$

Take one further example,

$$u_{n+2} + u_{n+1} - 20u_n = 28(2^n) \qquad \begin{aligned} u_0 &= 3 \\ u_1 &= -2 \end{aligned}$$

Step 1 Try the function $w_n = a(2^n)$ in the equation

$$a(2^{n+2}) + a(2^{n+1}) - 20a(2^n) = 28(2^n)$$

Dividing by 2^n, we have

$$\begin{aligned} 4a + 2a - 20a &= 28 \\ -14a &= 28 \\ a &= -2 \qquad w_n = -2(2^n) \end{aligned}$$

Step 2 Try $v_n = cr^n$ to make the left-hand side zero:

$$cr^{n+2} + cr^{n+1} - 20cr^n = 0$$
$$r^2 + r - 20 = 0$$
$$(r + 5)(r - 4) = 0$$
$$r = -5 \qquad r = 4 \qquad v_n = c_1(-5)^n + c_2(4)^n$$

Step 3 Combine the two solutions:

$$u_n = c_1(-5)^n + c_2(4)^n - 2(2^n)$$

Step 4 Find the constants by using $u_0 = 3$ and $u_1 = -2$:

$$\begin{aligned} 3 &= c_1 + c_2 - 2 & c_1 + c_2 &= 5 \\ -2 &= -5c_1 + 4c_2 - 4 & -5c_1 + 4c_2 &= 2 \end{aligned}$$

Solving, we obtain $c_1 = 2$ and $c_2 = 3$ and the final answer is

$$u_n = 2(-5)^n + 3(4)^n - 2(2)^n$$

EXERCISES

1. Solve the following recurrences:
   (a) $u_{n+2} - 4u_{n+1} + 3u_n = 0;\ u_0 = 2,\ u_1 = 3$
   (b) $u_{n+2} - 4u_n = 0;\ u_0 = 1,\ u_1 = 4$
2. Solve the following recurrences:
   (a) $u_{n+2} - 4u_{n+1} + 3u_n = n + 1;\ u_0 = 3,\ u_1 = 5$
   (b) $u_{n+2} - 4u_n = 3^n;\ u_0 = 2,\ u_1 = 2$
3. Solve the following recurrences:
   (a) $u_{n+2} + 4u_{n+1} - 5u_n = 0;\ u_0 = 2,\ u_1 = 6$
   (b) $u_{n+2} + 4u_{n+1} - 5u_n = 270(10^n);\ u_0 = 0,\ u_1 = 4$
4. Solve the following recurrences:
   (a) $u_{n+2} - u_{n+1} - 3u_n = 0;\ u_0 = 1,\ u_1 = 0$
   (b) $u_{n+2} - u_{n+1} - 3u_n = 9(4^n);\ u_0 = 2,\ u_1 = 4$
5. Find a general formula for the nth term of the sequences
   (a) 15, 30, 50, 75, 105, 140, . . . ;
   (b) 21, 27, 30, 30, 27, 21, ...
6. The Fibonacci numbers 1, 1, 2, 3, 5, 8, 13, 21, . . . are generated by the rule $u_{n+2} = u_{n+1} + u_n$, $u_0 = 1$, $u_1 = 1$. Find a formula for the nth Fibonacci number.

7. The numbers $1, 1, -3, -27, 135, \ldots$ are generated by the rule $u_{n+2} = 6u_{n+1} - 9u_n$, $u_0 = 1$, $u_1 = 1$. If you try the solution $u_n = cr^n$, it turns out that $r = 3$ is a double root of the equation and hence only one arbitrary constant occurs in the answer. Show that $u_n = cn(3^n)$ is another solution. Find the general formula for u_n.

7

Combinatorics

7.1 The Multiplication Principle

In this section we shall try to answer some problems involving counting. Suppose, for example, that a student must study one science and one foreign language to complete his degree requirements. He may choose biology, chemistry or physics as the science, and French, Spanish, German or Russian as the language. How many options does he have? All the possibilities can be exposed as shown below:

	French (F)	Spanish (S)	German (G)	Russian (R)
Biology (B)	(B, F)	(B, S)	(B, G)	(B, R)
Chemistry (C)	(C, F)	(C, S)	(C, G)	(C, R)
Physics (P)	(P, F)	(P, S)	(P, G)	(P, R)

Here we have listed all the possible pairs and see that they are twelve in number. In general, suppose an operation requires two actions. If there are n choices for the first action and m choices for the second action, then the display of possibilities would be an $n \times m$ table similar to the 3×4 table above. Clearly this display would have mn entries. This counting rule is called the *multiplication principle.*

Let us try to apply this principle to some simple examples. How many two-digit numbers between 30 and 70 can be formed using the digits 2, 3, 4, 5, 6, 7, 8? For the first digit we may choose 3, 4, 5 or 6. Therefore, there are four ways to pick the first digit. For the second digit we may choose any of the seven numerals 2, 3, 4, 5, 6, 7, 8. Thus, the total number of possible numbers is $4 \times 7 = 28$. How many may be formed if we may not use the same digit twice? In this case, we may choose the first digit in four ways. Whichever of the four digits is used, there remain six ways to choose the second digit; thus the total is $4 \times 6 = 24$. Now this is not quite the same as the two other examples. In counting the possible course options, we had four choices of language for each choice of a science, and we had a 3×4 table. In this example, for each choice of first digit we have six choices for the second digit, but not the same six choices in each case. In fact, we are really using multiplication as a shortcut for addition. That is, we are saying that beginning with 3 there are six possibilities:

$$32, 34, 35, 36, 37, 38$$

Beginning with 4 there are six possibilities:

$$42, 43, 45, 46, 47, 48$$

and so on for numbers beginning with 5 or 6. Nevertheless, the principle continues to apply because for each of the four possible first choices, there are six possible second choices.

On the other hand, suppose we wished to count only the odd numbers possible. Then, beginning with 3 we could form only 35 and 37, that is, two numbers. Beginning with 4 we have three numbers 43, 45 and 47; beginning with 5, 53 and 57; and beginning with 6, 63, 65 and 67. Thus, the total number of possibilities is ten. The procedure just applied is usually referred to as the *addition principle* and says simply that if A and B are disjoint sets, then the number of elements in $A \cup B$ is the sum of the number of elements in A and the number in B. Here we have automatically extended this principle to four disjoint sets, namely,

$$A = \text{set of odd numbers beginning with 3}$$

$$B = \text{set of odd numbers beginning with 4}$$

$$C = \text{set of odd numbers beginning with 5}$$
$$D = \text{set of odd numbers beginning with 6}$$

The multiplication principle can also be extended to more than two actions. Thus, the number of three-digit numbers between 300 and 600 that can be formed using 2, 3, 4, 5, 6, 7, 8 without repeating any digit is found as follows: The first digit may be chosen in three ways (3, 4 or 5). The second digit may now be chosen in six ways. Thus, the first two digits may be chosen in $3 \times 6 = 18$ ways. Now, the third digit may be chosen in five ways (since two digits have been used). Thus, the three digits may be chosen in $18 \times 5 = 3 \times 6 \times 5 = 90$ ways.

Sometimes we can combine the multiplication and addition principles. Thus, in the preceding example, count the number of odd three-digit numbers. In this case, it is easier to count by picking the second digit last. So, for the first digit we may select 3, 4 or 5. If the first digit is 3, then the last digit may be chosen in two ways (5 or 7) and the second digit may be chosen as any of the five remaining digits. Hence, we may choose $2 \times 5 = 10$ numbers starting with 3. If the first digit is 4, we may pick the last digit in three ways (3, 5 or 7) and the second digit in five ways. Hence, there are $3 \times 5 = 15$ numbers starting with 4. If the first digit is 5, then there are again $2 \times 5 = 10$ possibilities. Hence, the total number of possibilities is $10 + 15 + 10 = 35$.

One of the standard applications of the multiplication principle is in counting *permutations*. A permutation is simply a rearrangement. How many permutations (i.e., rearrangements) are there of the letters *CAT*? We can write them down:

$$CAT, CTA; TAC, TCA; ACT, ATC$$

six permutations. Note the count: The first letter can be chosen in three ways; the second in two ways and then the last can be chosen in only one way. Hence, we may count by the multiplication principle, $3 \times 2 \times 1 = 6$.

In general, the number of ways of "permuting" or rearranging n objects may be found by the multiplication principle. If we have n objects, we have n choices for the first position, we then have $n - 1$ choices for the second, $n - 2$ choices for the third and so on. Thus, the number of permutations of n objects is $n(n - 1)(n - 2) \ldots (3)(2)(1)$ which we shall call *n factorial*, symbolized $n!$. Thus

$$3! = 3 \cdot 2 \cdot 1 = 6$$
$$4! = 4 \cdot 3 \cdot 2 \cdot 1 = 24$$
$$5! = 5 \cdot 4 \cdot 3 \cdot 2 \cdot 1 = 120$$

Notice that $4! = 4(3!)$, $5! = 5(4!)$ or, in general, $n! = n(n - 1)!$ In order for this to be true, we define $0! = 1$. In summary, the number of possible orderings of n objects is $n! = n(n - 1) \ldots (3)(2)(1)$ and $0! = 1$ by definition.

EXERCISES

1. A club has twelve members. In how many ways may a president and a vice-president be selected?
2. On a dinner menu there is a choice of five entrees, three soups and four desserts. How many different complete dinners may be selected?
3. A club consists of seven men and ten women. In how many ways may a president and a vice-president be selected if the president must be a woman? How many if the president must be a woman and the vice-president a man?
4. In the club described in Exercise 3, a president, a vice-president and a secretary are to be selected. In how many ways can this be done if the president must be a man? If the president and vice-president must be of opposite sex? If exactly one officer is a man?
5. A class consists of six boys and eight girls. In how many ways may the class line up?
6. For the class in Exercise 5, how many lineups are possible if a boy must be at the head of the line? If a boy must be at the foot of the line and a girl at the head?
7. For the class in Exercise 5, how many lineups are possible if all the girls must be together and all the boys must be together?
8. How many arrangements of the letters in the word *ambidextrous* are possible? How many if the first letter must be A? How many if the first letter must be a vowel? How many if the first and last letters must be vowels?
9. A license plate consists of a letter followed by four numerals. How many license plates can be formed?
10. A license plate consists of four numbers and one letter. How many possible plates are there? How many if the first symbol cannot be zero or the letter O?
11. What is the remainder when $12! + 1$ is divided by 9?

12. There are three airlines flying between New York and Miami, four between Miami and Los Angeles and five between Los Angeles and New York. How many triangular round trips beginning and ending in New York are possible?
13. How many arrangements of ten men and ten women in a line are possible if men and women must alternate?

7.2 Combinations and Permutations

Let us consider the following type of problem. Given three letters *CAT*, in how many ways may we select a pair without regard to order? We may take *CA*, *CT* or *AT*. The pair *AC* is for this purpose the same as *CA*. The answer is then three. We shall use the symbol $\binom{3}{2}$ to mean the number of ways of selecting two objects out of three. In general, $\binom{n}{r}$ stands for the number of ways of selecting r objects out of n. Thus, if one were asked how many five-member committees can be selected from a group of twelve people, the answer would be $\binom{12}{5}$. Unfortunately, although we know that $\binom{3}{2} = 3$, we do not know what $\binom{12}{5}$ equals. Let us find out.

To begin, let us note that $\binom{12}{1}$ is the number of ways of picking one item out of twelve and that number is certainly 12. In fact, for any n,

$$\binom{n}{1} = n$$

Now returning to our committee, suppose we wish to have a committee of five, of which one person is the chairman. We could think of this as a two-stage operation. First, we select a chairman; there are twelve ways to do this. Then the chairman selects the other four members out of the remaining eleven people, and there are $\binom{11}{4}$ ways to do this. Thus, the total number of ways of selecting a committee with chairman is $12\binom{11}{4}$. Alternately, we could pick the five members of the committee (there are $\binom{12}{5}$ ways to do this) and have them select their own chairman (which can be done in five ways). Thus, the total number of ways of picking the committee is $5\binom{12}{5}$. We thus have the relationship

$$12\binom{11}{4} = 5\binom{12}{5}$$

In general, for picking a committee of size r with a chairman from a group of size n, we have

$$n\binom{n-1}{r-1} = r\binom{n}{r}$$

Ways to pick chairman	Ways for chairman to pick other members	Ways for members to pick chairman	Ways to pick members

We may rewrite this as

$$\binom{n}{r} = \frac{n}{r}\binom{n-1}{r-1}$$

Therefore,

$$\begin{aligned}\binom{12}{5} &= \frac{12}{5}\binom{11}{4}\\ &= \frac{12}{5}\cdot\frac{11}{4}\binom{10}{3}\\ &= \frac{12}{5}\cdot\frac{11}{4}\cdot\frac{10}{3}\binom{9}{2}\\ &= \frac{12}{5}\cdot\frac{11}{4}\cdot\frac{10}{3}\cdot\frac{9}{2}\binom{8}{1}\end{aligned}$$

Since $\binom{n}{1} = n$, $\binom{8}{1} = 8$, and we have

$$\binom{12}{5} = \frac{12\cdot 11\cdot 10\cdot 9\cdot 8}{5\cdot 4\cdot 3\cdot 2\cdot 1} = 792$$

There are 792 possible committees of size five that can be chosen from a group of twelve people.

In general,

$$\binom{n}{r} = \frac{n}{r}\binom{n-1}{r-1} = \frac{n(n-1)}{r(r-1)}\binom{n-2}{r-2} = \cdots$$
$$= \frac{n(n-1)(n-2)\ldots(n-r+1)}{r(r-1)(r-2)\ldots(1)}$$

Notice that the denominator of this fraction is $r!$ and the numerator is the product of r numbers counting down from n. Also, notice that $12! = 12\cdot 11\cdot 10\cdot 9\cdot 8\cdot 7!$, so we could write

$$\binom{12}{5} = \frac{12\cdot 11\cdot 10\cdot 9\cdot 8\cdot 7!}{5!7!} = \frac{12!}{5!7!}$$

and, in general,

$$\binom{n}{r} = \frac{n!}{r!(n-r)!}$$

The symbol representing the number of ways of selecting r objects out of a group size n is called *the number of combinations of n objects taken r at a time*. It may be interpreted as the number of different subsets of r objects which may be chosen from a set of n objects. Interestingly, there is yet another interpretation. Suppose we have r red balls and $n - r$ white balls to be arranged in a row. How many distinguishable arrangements are possible? Two arrangements obtained by simply interchanging the locations of two balls of the same color would not be distinguishable. Thus, we may think of the problem as follows: There are n possible locations. Of them, r must be occupied by red balls. Hence, the number of arrangements is simply the number of ways of selecting r locations out of the n possible ones.

Thus, for example, the number of distinguishable arrangements of five A's and six B's in a row is

$$\binom{11}{5} = \frac{11!}{5!6!} = \frac{11\cdot 10\cdot 9\cdot 8\cdot 7}{5\cdot 4\cdot 3\cdot 2\cdot 1} = 462$$

This reasoning may be extended. For example, how many distinguishable arrangements are there of the letters AAA, BB, CC? First, we pick three locations for the A's. This may be done in $\binom{7}{3}$ ways. Next we pick two of the remaining four locations for the B's. This may be done in $\binom{4}{2}$ ways.

The C locations are now forced on us. Thus, by the multiplication principle, the number of possible arrangements is

$$\binom{7}{3}\binom{4}{2} = \frac{7!}{3!4!} \cdot \frac{4!}{2!2!} = \frac{7!}{3!2!2!}$$

which is frequently written $\binom{7}{3, 2, 2}$ and is called a *multinomial coefficient.* Incidentally, $\binom{7}{3, 2, 2} = 210$.

This could again be generalized. A professor decided to give four A's, three B's, seven C's, one D, and five F's to a class of twenty students. The number of possible ways of doing this would be

$$\binom{20}{4, 3, 7, 1, 5} = \frac{20!}{4!3!7!1!5!}$$

which is about 28 billion ways. The reasoning for this is as shown.

$$\text{Ways to pick } A \text{ students, 4 out of 20} = \binom{20}{4}$$

$$\text{Ways to pick } B \text{ students, 3 out of remaining 16} = \binom{16}{3}$$

$$\text{Ways to pick } C \text{ students, 7 out of remaining 13} = \binom{13}{7}$$

$$\text{Ways to pick } D \text{ students, 1 out of remaining 6} = \binom{6}{1}$$

$$\text{Ways to pick } F \text{ students, 5 out of remaining 5} = \binom{5}{5}$$

$$\text{Total number} = \binom{20}{4}\binom{16}{3}\binom{13}{7}\binom{6}{1}\binom{5}{5}$$

which reduces to the previous result. Notice that $\binom{5}{5} = \frac{5!}{0!5!} = 1$.

It is not hard to see that, in general, if there are k different kinds of objects, r_1 of the first kind, r_2 of the second kind, etc., and if the total number of objects is n, then the number of distinguishable arrangements is

$$\binom{n}{r_1, r_2, \ldots, r_k} = \frac{n!}{r_1!r_2! \ldots r_k!}$$

When all the objects are different, $r_1 = r_2 = \cdots = r_k = 1$ and this expression reduces to $n!$, the number of permutations that we had earlier.

As a last type of count, let us consider the problem of selecting groups of r objects out of n where order counts. For example, how many different ways are there of selecting officers, president, vice-president, secretary and treasurer, from a club with twelve members? The simplest way to look at this is to say, There are twelve ways to pick a president, eleven ways to then pick a vice-president, ten ways to then pick a secretary and nine ways to pick a treasurer. Thus, the total is $12 \cdot 11 \cdot 10 \cdot 9 = 11{,}880$. Another way to count this is to say that there are $\binom{12}{4}$ ways to pick the officers and then $4!$ ways to arrange them by rank. This yields

$$\binom{12}{4} 4! = \frac{12!}{8!4!} 4! = \frac{12!}{8!} = (12)(11)(10)(9)$$

again! This number is referred to as the *number of permutations of twelve objects taken four at a time*, and is symbolized $(12)_4$. In general,

$$(n)_r = \binom{n}{r} r!$$

If you prefer, you may write this

$$(n)_r = \frac{n!}{(n-r)!} \quad \text{or} \quad (n)_r = n(n-1)(n-2)\ldots(n-r+1)$$

However, this is a case where it is easier to do it than to write it.

EXERCISES

1. A rummy hand consists of seven cards dealt from a 52-card deck. How many different rummy hands can a player be dealt?
2. An automobile dealer wants to arrange a Toyota, a Ford, a Plymouth, and a Batmobile along a straight curb in order to take publicity pictures. In how many distinct ways can the cars be arranged for the pictures?
3. A student's major department offers six different courses, while his minor department offers five different courses. In how many different ways can a student choose a program consisting of three courses

from his major department and two courses from his minor department?

4. Four of six people will be chosen to ride in a motorboat. One of the six owns the boat and must drive it. Find the number of possible seating arrangements for the ride.

5. A woman has five sons and four daughters. She chooses four sons and two daughters to help with the housework. In how many different ways can she do this?

6. A social studies teacher has a list of five different history books and seven different geography books, from which each student must choose three history books and five geography books to read. In how many ways can a student choose his reading assignment?

7. (a) In how many distinct ways can the letters in the word *calculus* be arranged?
   (b) How many arrangements can be formed from the letters of the word *Tennessee*, taken all at a time?
   (c) Find the number of different permutations which can be formed from the letters of the word *Cincinnati*, taken all at a time.

8. From a group of nine men and seven women a committee of five is to be chosen. Find the number of such committees containing exactly three men.

9. In how many different ways can six books be placed on a shelf so that two particular books are next to each other?

10. How many committees, each consisting of four men and three women, can be selected from a group of six men and seven women?

11. How many different permutations of the letters *ABCdefg* can be formed if each must begin and end with a capital letter and all the letters are used in each permutation?

12. How many committees of five can be formed from a group of six Democrats and three Republicans if each committee is to have at least four Democrats?

13. A signal is formed by placing nine flags in a vertical line on a flagpole. How many different signals of nine flags each can be formed from four red, three blue, and two white flags?

14. (a) How many different permutations can be formed from the letters of the word *perennial*, taken all at a time?
    (b) How many different permutations can be formed from the letters of the word *assassins*, taken all at a time?

(c) How many distinct permutations can be made from the letters of the word *vacation*, if they are all used every time?

7.3 Binomial Formula

Let us consider some properties of the combination symbol $\binom{n}{r}$. To begin, we state the following.

Property 1 $\binom{n}{r} = \binom{n}{n-r}$.

This follows from directly substituting $n - r$ for r in the formula. Thus

$$\binom{n}{n-r} = \frac{n!}{[n-(n-r)]!(n-r)!} = \frac{n!}{r!(n-r)!} = \binom{n}{r}$$

Alternatively, one could simply realize that every time a group of size r is selected from n objects, there remains a group of size $n - r$. Thus, there are just as many ways of selecting size r groups as size $n - r$ groups.

Property 2 $\binom{n}{0} = \binom{n}{n} = 1$.

This, of course, is simply obtained by substituting $r = 0$ and remembering that $0! = 1$.

Property 3 $\binom{n}{r-1} + \binom{n-1}{r-1} = \binom{n}{r}$.

This could be proved by algebra but there is a nice combinatoric argument for it. Suppose you are one of a group of n people from which a committee of size r is to be chosen. The total number of possible committees is $\binom{n}{r}$. These can be divided into those of which you are a member and those of which you are not a member. There are $\binom{n-1}{r}$ ways of choosing the committee when you are not a member since all r people must be chosen from the $n - 1$ others. If you are a member, then you need select only $r - 1$ other members and the number of ways of doing this is $\binom{n-1}{r-1}$.

Since these two possibilities exhaust the ways in which committees can be formed, we have

$$\binom{n}{r} = \binom{n-1}{r-1} + \binom{n-1}{r}$$

Using this relationship and $\binom{n}{0} = \binom{n}{n} = 1$, we may generate these coefficients via the so-called Pascal's triangle shown below:

$$\begin{array}{ccccccccccc l}
 & & & & & 1 & & & & & & \text{Row 0} \\
 & & & & 1 & & 1 & & & & & \text{Row 1} \\
 & & & 1 & & 2 & & 1 & & & & \text{Row 2} \\
 & & 1 & & 3 & & 3 & & 1 & & & \text{Row 3} \\
 & 1 & & 4 & & 6 & & 4 & & 1 & & \text{Row 4} \\
1 & & 5 & & 10 & & 10 & & 5 & & 1 & \text{Row 5} \\
 & & & & & \vdots & & & & & &
\end{array}$$

In each row, the first and last numbers are 1 and each other number is the sum of the two adjacently above it. Now, the fourth row gives $\binom{3}{r}$, $\binom{3}{0} = 1$, $\binom{3}{1} = 3$, $\binom{3}{2} = 3$, $\binom{3}{3} = 1$. The fourth row is

$$\binom{4}{0} = 1 \qquad \binom{4}{1} = \binom{3}{0} + \binom{3}{1} = 1 + 3 = 4$$

$$\binom{4}{2} = \binom{3}{1} + \binom{3}{2} = 3 + 3 = 6$$

and so on. Furthermore, we can use these numbers to generate the expansion of $(x + y)^n$. Since $x + y$ is a binomial, the formula for $(x + y)^n$ is called the *binomial* formula. We may generate this as follows: Consider $(1 + t)^n$:

$$(1 + t)^2 = 1 + 2t + t^2 \quad \text{and} \quad (1 + t)^3 = 1 + 3t + 3t^2 + t^3$$

can be worked out longhand. Notice that the coefficients of the powers of t are the entries in Row 2 and Row 3 of Pascal's triangle.

Now $(1 + t)^{n-1}$ must be a polynomial of order t^{n-1}; thus

$$(1 + t)^{n-1} = a_0 + a_1 t + a_2 t^2 + \cdots + a_{n-1} t^{n-1}$$

Multiplying by $(1 + t)$, we have

$$
\begin{aligned}
(1+t)^n &= (1+t)(a_0 + a_1 t + a_2 t^2 + \cdots + a_{n-1} t^{n-1}) \\
&= a_0 + a_1 t + a_2 t^2 + \cdots + a_{n-1} t^{n-1} \\
&\quad + a_0 t + a_1 t^2 + \cdots + a_{n-2} t^{n-1} + a_{n-1} t^n \\
&= a_0 + (a_1 + a_0) t + (a_2 + a_1) t^2 + \cdots \\
&\qquad\qquad + (a_{n-1} + a_{n-2}) t^{n-1} + a_{n-1} t^n
\end{aligned}
$$

Thus, we see that the coefficients of $(1 + t)^n$ start with a_0 and end with a_{n-1} unchanged for each n. Since for $n = 2$ and 3 the first and last are both 1, that must be so for all n. Furthermore, the coefficients of the powers of t in $(1 + t)^n$ are the sums of adjacent pairs of coefficients from the expansion of $(1 + t)^{n-1}$. Starting from $n = 2$, we can then generate these so-called binomial coefficients via Pascal's triangle exactly as we did $\binom{n}{r}$. We start with the same values for $n = 2$ and $n = 3$. Hence, they must be the identical numbers. Thus, we may write

$$(1+t)^n = \binom{n}{0} + \binom{n}{1} t + \binom{n}{2} t^2 + \cdots + \binom{n}{n} t^n$$

Letting $t = y/x$, we have $x^n(1 + t)^n$ is

$$
\begin{aligned}
x^n\left(1+\frac{y}{x}\right)^n &= x^n\left[\binom{n}{0} + \binom{n}{1}\frac{y}{x} + \binom{n}{2}\frac{y^2}{x^2} + \cdots + \binom{n}{n}\frac{y^n}{x^n}\right] \\
(x+y)^n &= \binom{n}{0} x^n + \binom{n}{1} x^{n-1} y + \binom{n}{2} x^{n-2} y^2 + \cdots + \binom{n}{n} y^n
\end{aligned}
$$

This last is the binomial formula. For example, from Pascal's triangle, the coefficients for $(x + y)^4$ are 1, 4, 6, 4, 1 and the expansion is

$$(x+y)^4 = x^4 + 4x^3y + 6x^2y^2 + 4xy^3 + y^4$$

More difficult problems are also possible. Thus,

$$(2x - 3y)^4 = 16x^4 - 96x^3y + 216x^2y^2 - 216xy^3 + 81y^4$$

is obtained by substituting $2x$ for x and $-3y$ for y in the formula.

Of course, by using this formula, one could find a particular term in the expansion without computing all the terms. For example, find the fifth term of $(x + 2y)^{13}$. The first term involves $x^{13}y^0$; second, $x^{12}y^1$; and the fifth, x^9y^4. Notice that the power of x and the power of y must sum to n (in this case 13). Therefore, the fifth term is

$$\binom{13}{4} x^9(2y)^4 = 715x^9(16y^4) = 11{,}440x^9y^4$$

Similarly, the coefficient of x^3y^8 in the expansion of $(2x - y)^{11}$ may be found by realizing that to have y^8 the term must be

$$\binom{11}{8}(2x)^3(-y)^8 = \binom{11}{8} 2^3(-1)^8x^3y^8$$
$$= 528x^3y^8$$

Hence, the coefficient is 528.

EXERCISES

1. Write out Rows 0 to 12 of Pascal's triangle.
2. Use the result of Exercise 1 to compute $(x + y)^{10}$.
3. Use the result of Exercise 1 to compute $(2x - y/2)^6$.
4. Compute $(x - 1/x)^7$.
5. Find the coefficient of x^2y^5 in the expansion of $(3x - y/3)^7$.
6. Find the coefficient of x^6y^3 in the expansion of $(x^2 + y)^6$.
7. Find the coefficient of x in the expansion of $(x^2 - 2/x)^8$.
8. Find and simplify the fifth term only in the expansion of $(2x - 3y)^9$.
9. Find and simplify the fifth term only in the expansion of $(2x - y^2/2)^6$.
10. Find and simplify the sixth term in the expansion of $(a^2 + 1/a^2)^{10}$.
11. Write the first four terms of the expansion of $(x/y - y/2x^2)^8$ and simplify each term.
12. Find and simplify the middle term in the expansion of $(y^2 - 1/2)^8$.
13. Write and simplify the middle term *only* in the expansion of $(2/x - x^2/2)^8$.
14. Write and simplify the middle term *only* in the expansion of $(a/x^2 + x/a^2)^{10}$.

7.4 Counting and Probabilities

The most basic type of probability problem involves probability spaces having N elementary outcomes, all of which are equally likely. That is, S consists of N points, each having likelihood $1/N$. Now if $E \subset S$ is any event, then

$$P(E) = k(1/N)$$

where k is the number of elements of E. Since

$$k(1/N) = k/N$$

this is expressed by saying

$$P(E) = \frac{\text{number of outcomes favorable to } E}{\text{total number of possible outcomes}}$$

This is only valid when all N possible outcomes are equally likely.

This rule is relatively simple to apply when the counting of possibilities is simple to do. Thus, you are part of a class of twenty students, from which one will be selected to put a problem on the blackboard. What is the probability that you are chosen? In this case, $N = 20$, $k = 1$ and $P(\text{You}) = 1/20$. Now, suppose that three students will be chosen. What is the probability that you are one of the three? In this case $N = \binom{20}{3}$, the number of ways of choosing three out of twenty. The number of possible sets which do not include you is $\binom{19}{3}$. Thus,

$$P(\text{Not You}) = \binom{19}{3} \Big/ \binom{20}{3} = \frac{19 \cdot 18 \cdot 17}{3 \cdot 2 \cdot 1} \div \frac{20 \cdot 19 \cdot 18}{3 \cdot 2 \cdot 1} = \frac{17}{20}$$

and

$$P(\text{You}) = 1 - P(\text{Not You}) = 1 - \frac{17}{20} = \frac{3}{20}$$

Examples may be even more difficult. Suppose one is dealt a poker hand of five cards from a standard deck. What is the probability of having a "full house"? (A full house is three cards of one denomination and two cards of another denomination.) The number of possible poker hands is $\binom{52}{5}$. The number of possible full houses is more difficult to count. First, there are thirteen possible ways to choose the denomination for the threesome and $\binom{4}{3}$ to pick the three cards of that denomination. There are then twelve ways to choose the denomination of the pair and $\binom{4}{2}$ to pick the two cards of that denomination. Thus

$$P(\text{Full House}) = \frac{(13) \cdot \binom{4}{3} \cdot (12) \cdot \binom{4}{2}}{\binom{52}{5}} = \frac{6}{4{,}165}$$

or about one chance in 700.

A deceptively simple problem follows. There are tickets numbered 1, 2, 3, . . . , 1,000 in a lottery. Four tickets are drawn successively. What is the probability that they are in increasing order? Notice, we did not say in sequence. Thus, 3, 11, 78, 651 are in increasing order although not in sequence. In this case, whatever the four numbers be, there are 4! possible arrangements. Since only one such can be the proper order, the probability is $1/4! = 1/24$.

Consider another example. An assembly line produces 2,000 items of which 600 are faulty. Twenty items are selected at random. What is the probability that exactly five are defective? The number of possible drawings is $\binom{2{,}000}{20}$, since we must pick twenty out of 2,000. Letting X be the number of defectives, to have $X = 5$, we must pick five out of the 600 faulty items. This can be done in $\binom{600}{5}$ ways. We must pick the remaining fifteen out of the 1,400 good items. This can be done in $\binom{1{,}400}{15}$ ways. Thus, the number of favorable outcomes to the event "$X = 5$" is $\binom{600}{5} \cdot \binom{1{,}400}{15}$ and

$$P(X = 5) = \frac{\binom{600}{5} \cdot \binom{1{,}400}{15}}{\binom{2{,}000}{20}}$$

This would work out to be about 18%.

We have here an example of a so-called hypergeometric distribution. The population consists of N objects (in this case 2,000). There are D items of interest (in this case the 600 defectives). There are $N - D$ remaining items. From the population, n items are chosen and X is the number of items of interest. By the same reasoning as above, we obtain

$$P(X = k) = \frac{\binom{D}{k}\binom{N - D}{n - k}}{\binom{N}{n}}$$

Naturally, there are certain values of k which cannot occur. In the preceding example, even if we had drawn 700 items ($n = 700$), we could still not draw more than 600 defectives. However, if we tried to substitute $k = 625$

$$P(X = 625) = \frac{\binom{600}{625}\binom{1{,}400}{75}}{\binom{2{,}000}{700}}$$

we see the expression $\binom{600}{625}$. This is the number of ways of drawing 625 items out of 600, which is certainly nonsense. If we just interpret any nonsensical binomial coefficient to be zero, the hypergeometric formula will automatically prevent us from asking incorrect questions.

EXERCISES

1. Six married couples are seated at random at a straight banquet table. What is the probability that Mr. and Mrs. A sit next to each other?
2. What is the probability of drawing four-of-a-kind in a five-card poker hand?
3. What is the probability that a bridge hand of thirteen cards will have flat distribution (four cards of one suit and three of each other)?
4. (a) A committee of four persons is selected from among five women and six men. What is the probability that the committee will consist of three men and one woman?
   (b) A bag contains five white and three red balls. If three balls are taken together at random, what is the probability that two are white and one is red?
   (c) An urn contains seven red and four black balls. If three balls are drawn together, what is the probability that two are red and one is black?
5. (a) Box A contains 100 bulbs, sixty of which are defective, while Box B contains 100 bulbs ten of which are defective. A box is chosen at random and three bulbs are drawn at random from the box. Compute the probability that one bulb is defective.
   (b) If one defective bulb has been drawn, what is the probability that it came from Box B?

6. (a) A box contains five red, four white and three black balls. If two balls are drawn simultaneously, what is the probability that both are white?
   (b) What is the probability that at least two are white if four balls are drawn?
   (c) What is the probability that exactly two are white and two are red if five are drawn?

7. Five different books are to be placed randomly on a shelf. What is the probability that two certain books will be next to each other?
8. A box contains ten flashbulbs, two of which are defective. Four bulbs are drawn. Let X be the number of defectives. Find the $P(X = 0)$, $P(X = 1)$, $P(X = 2)$ and the expected value of X.
9. For the situation in Exercise 6, find the expected number of red balls drawn in three picks.
10. From a group of six Democrats and eight Republicans, five people are selected at random. What is the probability that a majority of the five are Republicans?
11. Six couples line up in a row. What is the probability that the line alternates men and women?
12. An urn contains four white and two black balls. Three balls are chosen at random. For every black ball drawn, one additional ball is drawn. What is the expected value of the total number of black balls drawn?

7.5 The Binomial Distribution

Let us consider another random variable closely associated with the hypergeometric. Suppose we have an experiment in which there are two possible outcomes, success and failure. For example, in flipping a coin we might call heads a success, and tails a failure. Suppose the probability of a success is p. The probability of a failure is $q = 1 - p$. Let us have n independent repetitions of the experiment. For each trial we may have success or failure. Let X be the number of successes. We shall use the symbol $b(k; n, p)$ to stand for the probability of exactly k successes, that is,

$$b(k; n, p) = P(X = k)$$

Let us first consider $P(X = 0) = b(0; n, p)$. To have $X = 0$, we must get a string of n failures. The probability of a failure on each trial is q. Since the trials are independent, we may write

$$\begin{aligned} b(0; n, p) = P(F_1, F_2, F_3, \ldots, F_n) &= P(F_1)P(F_2)P(F_3) \ldots P(F_n) \\ &= \underbrace{q \cdot q \cdot q \cdot \ldots \cdot q}_{n \text{ times}} \\ &= q^n \end{aligned}$$

where the subscripts on the F's indicate the number of the trial. Computing $b(1; n, p)$ we must realize that

$$\begin{aligned} P(X = 1) = {} & P(S_1, F_2, F_3, \ldots, F_n) \\ & + P(F_1, S_2, F_3, \ldots, F_n) \\ & \vdots \\ & + P(F_1, F_2, F_3, \ldots, S_n) \end{aligned}$$

That is, the one success may occur on any of the n trials. Again using independence,

$$P(S_1, F_2, F_3, \ldots, F_n) = P(S_1)P(F_2) \ldots P(F_n) = pq^{n-1}$$

and this is the same for all the strings having one success. There are exactly $\binom{n}{1} = n$ ways in which we may have exactly one success in n trials and hence

$$b(1; n, p) = \binom{n}{1} pq^{n-1}$$

Similarly, to compute $P(X = k)$ one must consider all possible strings of kS's and $(n - k)F$'s in a row. There are, of course, $\binom{n}{k}$ of these. Each one has the same probability as

$$S_1, S_2 \ldots S_k, F_{k+1} \ldots F_n$$

$$\begin{aligned} P(S_1, S_2 \ldots S_k, F_{k+1} \ldots F_n) = P(S_1)P(S_2) \ldots \\ P(S_k)P(F_{k+1}) \ldots P(F_n) = p^k q^{n-k} \end{aligned}$$

Thus,

$$b(k; n, p) = \binom{n}{k} p^k q^{n-k} \quad \text{for } k = 0, 1, 2, \ldots, n$$

This last expression is called the *binomial probability function*, because each probability is exactly the term in the binomial expansion of $(p + q)^n$. Since $p + q = 1$, it is easy to see that the sum of the probabilities is $1^n = 1$, as it must be.

Binomial probability problems may arise in a variety of guises. The most common involves the repetition of a simple experiment. Thus, a pair of dice is rolled eight times. What is the probability of getting exactly two 7's? On any roll, the probability of a seven is 1/6. Therefore, we have a binomial problem with $p = 1/6$, $q = 5/6$ and $n = 8$.

$$P(X = 2) = b\left(2; 8, \frac{1}{6}\right) = \binom{8}{2}\left(\frac{1}{6}\right)^2\left(\frac{5}{6}\right)^6 = 0.26$$

These probabilities cannot be easily computed and hence they have been tabulated. A short table of this distribution for $n = 1, 2, \ldots, 20$ is included in Appendix 3 of this book. Let us take another example. A professional basketball player makes 40% of his shots. Find the probability that out of ten shots he makes, at most, three. In this case, $p = 0.4$, $q = 0.6$, $n = 10$ and we want

$$P(X \leq 3) = b(0; 10, 0.4) + b(1; 10, 0.4) + b(2; 10, 0.4) + b(3; 10, 0.4)$$

Looking these up in the table, we find

$$P(X \leq 3) = 0.0060 + 0.0403 + 0.1209 + 0.2150 = 0.3844$$

Again, the probability that a student passes math is 0.9; find the probability that in a class of eighteen students, precisely sixteen pass. In this case

$$P(X = 16) = b(16; 18, 0.9)$$

A glance at the table shows that no p value over 0.5 is included. Hence, we must let Y be the number of *failures*. Then p will equal 0.1 and to have sixteen pass, we must have two fail; hence

$$P(Y = 2) = b(2; 18, 0.1) = 0.2835$$

In general, in order to use our tables, we must always choose as a "success" the possibility which has the smaller probability, so that p will not exceed 1/2.

A standard case in which the binomial occurs is *sampling with replacement*. In this situation, a batch of items is to be sampled. Items are selected one by one at random and after each item is inspected, it is returned to the batch and may be chosen again. In this way, if we had, say, 100 items ten of which are defective, on each pick there are always ten defectives and ninety good items available to be chosen. Therefore, on each pick the probability of a defective is 0.1 regardless of what has gone before. If the items are not returned, we have sampling without replacement and we are dealing with the hypergeometric case. However, if in a hypergeometric case, the population is quite large and the sample is small, then for all practical purposes, it makes no difference whether we sample with or without replacement. For example, there are 10,000,000 people eligible to vote in New York State. Of these, 7,000,000 are registered. Twenty eligible voters are selected at random. What is the probability that fewer than seventeen have registered? Let X be the number *not* registered. We want the probability that X is 4 or more. It is easiest to write

$$P(X \geq 4) = 1 - [P(X = 0) + P(X = 1) + P(X = 2) + P(X = 3)]$$

Now actually X has hypergeometric distribution. But twenty voters, more or less, will not really change the fact that the probability of selecting a registered voter is 0.7 and an unregistered voter 0.3. Thus, we may treat X as a binomial with $n = 20$ and $p = 0.3$ and

$$\begin{aligned} P(X \geq 4) &= 1 - b(0; 20, 0.3) - b(1; 20, 0.3) - b(2; 20, 0.3) - b(3; 20, 0.3) \\ &= 1 - 0.0008 - 0.0068 - 0.0278 - 0.0716 = 0.8930 \end{aligned}$$

How does this answer strike you? Twenty eligibles are picked at random, there is almost a 90% probability that fewer than seventeen are registered. Let us consider. Seven out of every ten eligible people register to vote. Therefore, out of twenty people, we should expect fourteen to be registered, so it seems likely that the observed number would be less than seventeen.

We just said that we should expect fourteen people to be registered. Is that right? If we flip a fair coin 100 times, how many heads do you expect? If we roll a fair die sixty times, how many 1's do you expect to roll? You should have answered, "50 heads and ten 1's." In fact, what we shall prove is that in n independent trials with success probability p, the expected number of successes is np. In other words, if X is binomial, $E(X) = np$. We may prove this by proceeding from the definition. Remember, if X is a random variable and $P(X = k) = p(k)$, $k = 0, 1, 2, \ldots, n$; then

$$E(X) = 0 \cdot p(0) + 1 \cdot p(1) + 2 \cdot p(2) + \cdots + n \cdot p(n)$$

In the binomial case,

$$E(X) = 0\cdot b(0; n, p) + 1\cdot b(1; n, p) + 2\cdot b(2; n, p) + \cdots + n\cdot b(n; n, p)$$
$$= 0\binom{n}{0} q^n + 1\binom{n}{1} pq^{n-1} + 2\binom{n}{2} p^2q^{n-2} + \cdots + n\binom{n}{n} p^n$$

Earlier, we derived the binomial coefficients by noting that

$$\binom{n}{k} = \frac{n}{k}\binom{n-1}{k-1} \quad \text{or} \quad k\binom{n}{k} = n\binom{n-1}{k-1}$$

Thus,

$$1\binom{n}{1} = n\binom{n-1}{0}; 2\binom{n}{2} = n\binom{n-1}{1}; 3\binom{n}{3} = n\binom{n-1}{2};$$
$$\ldots n\binom{n}{n} = n\binom{n-1}{n-1}$$

Therefore,

$$E(X) = n\binom{n-1}{0} pq^{n-1} + n\binom{n-1}{1} p^2q^{n-2}$$
$$+ n\binom{n-1}{2} p^3q^{n-3} + \cdots + n\binom{n-1}{n-1} p^n$$

Factoring np from each term, we have

$$E(X) = np\left[\binom{n-1}{0} q^{n-1} + \binom{n-1}{1} pq^{n-2}\right.$$
$$\left. + \binom{n-1}{2} p^2q^{n-3} + \cdots + \binom{n-1}{n-1} p^{n-1}\right]$$
$$= np\cdot(p+q)^{n-1} = np(1) = np$$

By an almost identical proof, we could show that

$$E[(X)(X-1)] = n(n-1)p^2$$

and hence

$$n(n-1)p^2 = E(X^2 - X) = E(X^2) - E(X) = E(X^2) - np$$

So,

$$E(X^2) = n(n-1)p^2 + np$$

and

$$\begin{aligned}\mathrm{Var}(X) &= E(X^2) - [E(X)]^2 \\ &= n(n-1)p^2 + np - (np)^2 = np(1-p) = npq\end{aligned}$$

The algebra is a little tricky in that last line, but give it a try.

In summary, we may state the following theorem.

THEOREM 1 Let X be the number of successes in n independent trials for which the probability of a success is p on each trial. Then

$$P(X = k) = \binom{n}{k} p^k q^{n-k} \quad \text{for } k = 0, 1, 2, \ldots, n$$

$$E(X) = np \quad \text{and} \quad \mathrm{Var}(X) = npq$$

where $q = 1 - p$.

As a final example, consider the following. Suppose that one out of every five Bostonians can trace his ancestors to the *Mayflower*. In a sample of 100 Bostonians, we would expect $20 = 100 \cdot 1/5$ to be able to thus trace his ancestors. Calling X the number of such Bostonians in the sample of 100, then $\mathrm{Var}(X) = 100 \cdot 1/5 \cdot 4/5 = 16$ and $\mathrm{SD}(X) = 4$. According to Chebyschev's inequality, the probability that X falls more than three standard deviations from its mean is at most 1/9. That is, there is at most an 11% probability that in a sample of size 100, X will be fewer than 8 (20 − 12), or more than 32 (20 + 12). Another way of saying this is that there is at least an 89% probability that X falls between 8 and 32. In fact, it is virtually certain that this will happen, as we shall see in the next section.

EXERCISES

1. Let X be a binomial random variable with $n = 14$ and $p = 0.2$. Find $P(X = 2)$, $P(X \leq 2)$, $P(X > 11)$, $P(8 < X \leq 12)$.
2. A biased coin is flipped twelve times. The probability of a head on each flip is 0.4. Find the probabilities of
   (a) exactly six heads;
   (b) more than ten heads;
   (c) less than nine heads;
   (d) no more than two heads;
   (e) no less than eleven heads.

3. For X in Exercise 1, find $E(X)$ and $\text{Var}(X)$.

4. Find the expected number of heads in the experiment described in Exercise 2.

5. The probability that a student selected at random has taken math is 0.7. In a group of fifteen students, how many would you expect to have taken math? What is the probability that ten have taken math? Eleven? What is the most probable number?

6. The probability that a man is tall is 2/5. What is the probability that of six men chosen at random, at least four are tall? Let X be the number of tall men in the sample. Find $E(X)$ and $\text{Var}(X)$.

7. The probability that the Knickerbockers will beat the Bucks is 0.6. What is the probability that the Knicks will win a seven-game series from the Bucks in four straight games? What is the probability that they will win in five games? Seven games?

8. What is the probability that a World Series between evenly matched teams will go seven games?

9. In a certain city 30% of the people have brown hair, 20% have black hair, 50% have blond hair. Of fourteen people chosen at random from the town, what is the probability that at least three of them have black hair?

10. It is claimed that among all drivers whose cars are equipped with seat belts, 75% use them on long trips. What is the probability that among five cars (equipped with seat belts) which pass through a toll station on a turnpike, three of the drivers are using seat belts?

11. According to one authority, the expected number of seat belt wearers is eight out of ten people. What is the probability that at least twelve out of fourteen people wear seat belts?

12. According to one authority, 75% of all people wear seat belts. According to a second authority, 50% of all people wear seat belts. In a sample of twelve people, five wore seat belts. If originally you thought both authorities equally reliable, how do you now feel about the probabilities of each being correct? HINT: Use Bayes' theorem approach.

7.6 The Central Limit Theorem

Let us reconsider the voter registration example: the number of trials, $n = 20$; the probability of not being registered is $p = 0.3$; the expected number of nonregistrants is $E(X) = np = 20(0.3) = 6$ and $\text{SD}(X) =$

$\sqrt{npq} = \sqrt{(20)(0.3)(0.7)} = \sqrt{4.2} = 2.05$. According to Chebyschev's inequality, the probability that X falls within two standard deviations of its mean is at least 3/4, that is,

$$P[6 - 2(2.05) < X < 6 + 2(2.05)] \geq 0.75$$
$$P[1.9 < X < 10.1] \geq 0.75$$
$$P[X = 2, 3, 4, 5, 6, 7, 8, 9, 10] \geq 0.75$$

From the table $P(X = 2) + P(X = 3) + \cdots + P(X = 10) = 0.9753$. Thus, the true probability is over 97% that the binomial random variable will fall within two standard deviations of its expected value. Clearly, Chebyschev's inequality leaves something to be desired in terms of precision.

For the binomial random variable, if n and p are in one's tables, this is not a problem. Good tables for n up to 100 are available in many libraries, and computers can do the computations for larger n or nontabulated values of p. However, it turns out that this is really unnecessary. In the days before electronic computers, many mathematicians searched for a good approximation to the binomial distribution for large values of n. They were so successful in their quest that the need for large binomial tables is virtually nil. Their method involves approximating the binomial probabilities by the area under the familiar bell curve, which is also known as the *normal distribution* or *Gaussian distribution* function. The approximation rule, known as the central limit theorem, has been extended and amplified to cover far more than approximation to the binomial. However, for our purposes we shall only consider this first application.

We notice in Chebyschev's inequality that we measure

$$P(-k\sigma < X - \mu < k\sigma) \quad \text{or} \quad P\left(-k < \frac{X - \mu}{\sigma} < k\right)$$

and we arrive at an estimate which does not depend upon μ, σ or the specific probability distribution. This suggests that the expression

$$Z = \frac{X - \mu}{\sigma}$$

is somehow the right thing to work with. In fact, for $Z = (X - \mu)/\sigma$

$$E(Z) = \frac{1}{\sigma} E(X - \mu) = \frac{1}{\sigma} [E(X) - \mu]$$
$$= \frac{1}{\sigma} (\mu - \mu) = 0$$

Therefore, Z has mean 0 and hence $\text{Var}(Z) = E(Z^2)$:

$$\text{Var}(Z) = E(Z^2) = \frac{1}{\sigma^2} E(X - \mu)^2 = \frac{1}{\sigma^2} \text{Var}(X)$$
$$= \frac{1}{\sigma^2} \cdot \sigma^2 = 1$$

Therefore, Z has mean 0 and standard deviation 1. Such a random variable is called *standardized*. It is convenient to standardize when comparing random variables since their probability functions will then fit the same scale.

If Z is a standardized binomial random variable, then if n is large, $P(Z \leq b)$ is given approximately by the area under the bell curve to the left of b. We shall call this area $\Phi(b)$. (See figure 1.)

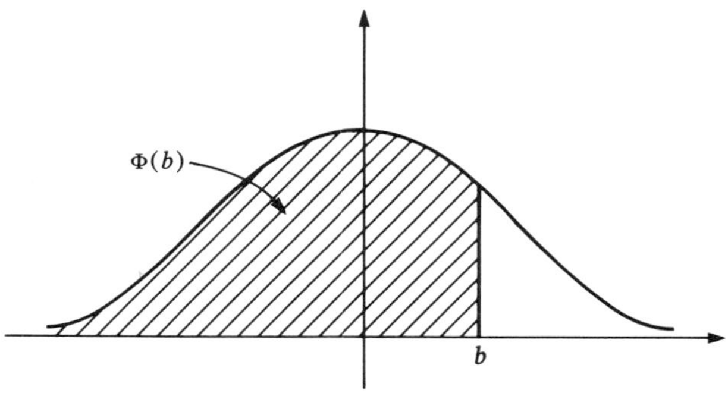

Figure 1

A table with which one may compute $\Phi(b)$ is included in Appendix 3. The values in the table actually give the area under the bell curve between 0 and b. The table is read down and then across. Thus, the area under the curve between 0 and 1.13 is found by reading down to $b = 1.1$ and then across over to the column headed 0.03. The entry in this spot is 0.3708. The area is illustrated in figure 2.

Actually, the bell curve is symmetric with area 1/2 on either side of 0. Hence, $\Phi(1.13) = P(Z \leq 1.13) =$ area between 0 and 1.13 + area to the left of $0 = 0.3708 + 0.5000 = 0.8708$. Thus, if b is positive, one finds the $P(Z \leq b)$ by adding 0.5 to the table entry for that value of b. To find $P(Z \leq -b)$ we also use symmetry as shown in figure 3. The shaded area is 0.3708, the same as between 0 and +1.13. Therefore, the left end tail has area $0.5000 - 0.3708 = 0.1292$. Hence

$$P(Z \leq -1.13) = 0.1292$$

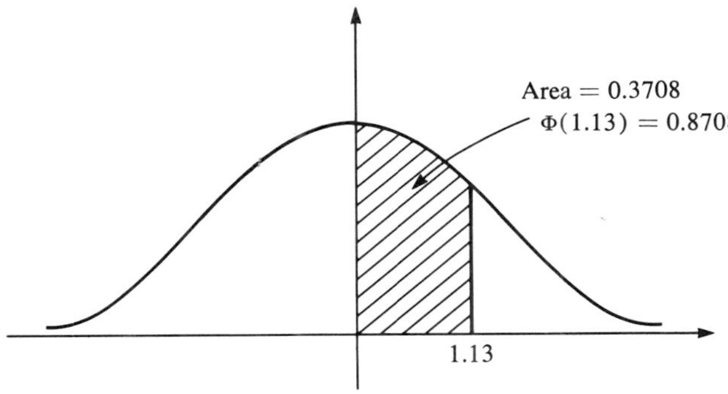

Figure 2

If $a < b$, we may write

$$\begin{aligned} P(Z \leq b) &= P(Z \leq a) + P(a < Z \leq b) \\ \Phi(b) &= \Phi(a) + P(a < Z \leq b) \\ P(a < Z \leq b) &= \Phi(b) - \Phi(a) \end{aligned}$$

For example, $P(-0.1 < Z \leq 1.26) = \Phi(1.26) - \Phi(-0.1)$:

$$\begin{aligned} \Phi(1.26) &= 0.3962 + 0.5 = 0.8962 \\ \Phi(-0.1) &= 0.5000 - 0.0398 = 0.4602 \end{aligned}$$

Therefore, $P(-0.1 < Z \leq 1.26) = 0.8962 - 0.4602 = 0.4360$

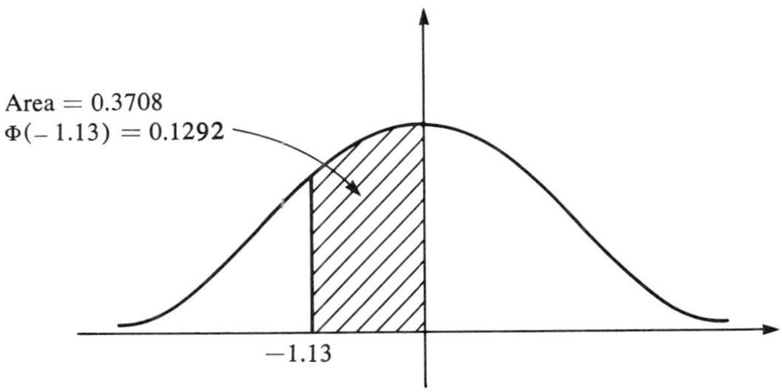

Figure 3

Now, to actually compute binomial probabilities, suppose X is binomial; then to find $P(X = k)$ write it in the form

$$P(X = k) = P[k - (1/2) < X \leq k + (1/2)]$$

This is a correct statement since the binomial cannot take on fractional values. Next, standardize

$$\begin{aligned} P(X = k) &= P[k - (1/2) < X \leq k + (1/2)] \\ &= P[k - (1/2) - \mu < X - \mu \leq k + (1/2) - \mu] \\ &= P\left[\frac{k - 1/2 - \mu}{\sigma} < \frac{X - \mu}{\sigma} \leq \frac{k + 1/2 - \mu}{\sigma}\right] \\ &= P\left[\frac{k - 1/2 - \mu}{\sigma} < Z \leq \frac{k + 1/2 - \mu}{\sigma}\right] \\ &= \Phi\left(\frac{k + 1/2 - \mu}{\sigma}\right) - \Phi\left(\frac{k - 1/2 - \mu}{\sigma}\right) \end{aligned}$$

where the last equality is only approximate. Remember that $\mu = np$ and $\sigma = \sqrt{npq}$.

Now let us try another problem. A fair coin is flipped sixteen times. The probability of getting eight heads may be looked up in the binomial table:

$$P(X = 8) = b(8; 16, 1/2) = 0.1964$$

Using the normal approximation,

$$\begin{aligned} \mu &= 16(1/2) = 8 \\ \sigma &= \sqrt{(16)(1/2)(1/2)} = 2 \end{aligned}$$

we obtain

$$\begin{aligned} P(X = 8) &= \Phi\left(\frac{8 + 1/2 - 8}{2}\right) - \left(\frac{8 - 1/2 - 8}{2}\right) \\ &= \Phi(0.25) - \Phi(-0.25) \\ \Phi(0.25) &= 0.5000 + 0.0987 = 0.5987 \\ \Phi(-0.25) &= 0.5000 - 0.0987 = 0.4013 \\ P(X = 8) &= 0.5487 - 0.4013 = 0.1974 \end{aligned}$$

with an error of about one part in two hundred!

Suppose we have $n = 18$, $p = 0.7$ and are to find $P(X \geq 14)$. Here, $P(X \geq 14) = P(X = 14) + P(X = 15) + \ldots$. Since we interpret

$$\begin{aligned} &P[X = 14] \quad \text{as} \quad P[13\tfrac{1}{2} < X \leq 14\tfrac{1}{2}] \\ &P[X = 15] \quad \text{as} \quad P[14\tfrac{1}{2} < X \leq 15\tfrac{1}{2}] \\ &\quad\vdots \end{aligned}$$

we shall write

$$\begin{aligned}P(X \geq 14) &= P[X > 13\tfrac{1}{2}] \\ &= 1 - P[X \leq 13\tfrac{1}{2}] \\ &= 1 - P\left[\frac{X - \mu}{\sigma} \leq \frac{13\tfrac{1}{2} - \mu}{\sigma}\right]\end{aligned}$$

Here $\mu = (18)(0.7) = 12.6$, $\sigma = \sqrt{(18)(0.7)(0.3)} = 1.94$ and

$$P[X \geq 14] = 1 - \Phi\left[\frac{13.5 - 12.6}{1.94}\right] = 1 - \Phi(0.46) = 0.3228$$

As a check, since X is binomial with $n = 18$ and $p = 0.7$, for $X \geq 14$, as earlier we must have $Y = 0, 1, 2, 3, 4$ where Y has $n = 18$ and $p = 0.3$. Thus

$$\begin{aligned}P[X \geq 14] &= b(0; 18, 0.3) + b(1; 18, 0.3) + b(2; 18, 0.3) \\ &\qquad + b(3; 18, 0.3) + b(4; 18, 0.3) \\ &= 0.0016 + 0.0126 + 0.0458 + 0.1046 + 0.1681 = 0.3327\end{aligned}$$

This time the approximation is not quite as good but remember, it is only intended to be used for large values of n.

EXERCISES

1. Find $\Phi(b)$ for $b = 0.2, 1.31, -2.2, -1.36$.
2. Find b such that $\Phi(b) = 0.6915, 0.8438, 0.9357, 0.0548, 0.0764, 0.4880$.
3. If Z is a standardized binomial variable with large n, find
   (a) $P(1 < Z \leq 2.5)$;
   (b) $P(-1 < Z \leq 2.5)$;
   (c) $P(-2.2 < Z \leq 2.36)$;
   (d) $P(-3 < Z \leq 3)$.
4. If Z is a standardized binomial variable with large n, find
   (a) $P(Z > -1.3)$;
   (b) $P(Z > 1.27)$;
   (c) $P(Z \leq 1.12)$;
   (d) $P(Z \leq -1.76)$.

5. Let X be a random variable with $n = 16$ and $p = 0.2$. Find
   (a) the exact probability that X falls within two standard deviations of its mean (use the binomial tables);
   (b) the Chebyschev bound for the probability of this event;
   (c) the normal approximation to the probability of this event.
6. A fair coin is flipped 100 times. (a) Find, approximately, the probability of getting exactly fifty heads. Find the probability of getting between 45 and 55 heads. (b) If the coin is flipped 10,000 times, find the probability of 5,000 heads and the probability that the number of heads is between 4,500 and 5,500. Interpret your results (between means not including the end points).
7. Repeat Exercise 6, only this time compute the probability that the fraction heads is between 0.49 and 0.51 for 100 flips, 10,000 flips and 1,000,000 flips. Interpret your results.
8. One-third of all television viewers watch "All in the Family." If 1,800 viewers are sampled at random, what is the probability that fewer than 650 watch that show? What is the probability that between 575 and 660 people watch that show?
9. A multiple choice test has fifty questions, each having four possible answers. What is the probability that a student guessing at random will score over 35% on the test?
10. (a) Find b, so that $\Phi(b) = 0.99$.
    (b) The probability that a gunner hits his target is 0.6. If he fires 96 shots, what is the probability that he makes less than 60 hits? 65 hits? 70 hits? For what number of hits is the probability 99%? 95%?
11. The probability that the Orioles will win a baseball game is 0.6. What is the probability that they will win over 100 games in a 162-game season?
12. A tennis critic says that Rod Laver has a 2/3 chance of winning a given point from Arthur Ashe. A record of 120 tie-breaker points between the two shows that Laver has won one-half. What do you think of the critic's judgment?
13. At the end of 7.5, we said that out of 100 Bostonians, it was virtually certain that between 8 and 32 of them could trace their ancestry back to the *Mayflower*. Show by the methods of this section that this is true.

8

Determinants

8.1 Systems of Two Equations in Two Unknowns

Let us consider the most general case of two equations in two unknowns:

$$\begin{aligned} ax + by &= e \\ cx + dy &= f \end{aligned}$$

which could also be written in the form

$$\mathbf{AX} = \mathbf{B}$$

where

$$\mathbf{A} = \begin{pmatrix} a & b \\ c & d \end{pmatrix} \qquad \mathbf{X} = \begin{pmatrix} x \\ y \end{pmatrix} \qquad \mathbf{B} = \begin{pmatrix} e \\ f \end{pmatrix}$$

We may attempt to solve this system by eliminating x and solving for y. To do so, multiply the first equation by c and the second by a which yields

$$acx + bcy = ce$$
$$acx + ady = af$$

Subtracting the first equation from the second yields

$$(ad - bc)y = af - ce$$
$$y = \frac{af - ce}{ad - bc} \quad \text{if } ad - bc \neq 0$$

Substituting in the first equation, we obtain

$$ax + \frac{b(af - ce)}{ad - bc} = e$$
$$ax = \frac{-baf + bce + ead - ebc}{ad - bc} = \frac{-baf + ead}{ad - bc}$$

Dividing by a yields

$$x = \frac{ed - bf}{ad - bc} \quad \text{if } ad - bc \neq 0$$

Notice that it is crucial that $ad - bc \neq 0$ if we are to have a unique solution to the system. This expression $ad - bc$ is called the *determinant* of the system or the determinant of the matrix $\mathbf{A}$, symbolized "det $\mathbf{A}$." Thus if

$$\mathbf{A} = \begin{pmatrix} 1 & 3 \\ 2 & 6 \end{pmatrix}$$

then

$$\det \mathbf{A} = \det \begin{pmatrix} 1 & 3 \\ 2 & 6 \end{pmatrix} = (1)(6) - (2)(3) = 0$$

and hence we know that

$$x + 3y = 5$$
$$2x + 6y = 7$$

has no unique solution. This pair of equations is, in fact, an incompatible set. The system

$$x + 3y = 3$$
$$2x + 6y = 6$$

also has no unique solution. But in this case the pair is redundant and we have an entire class of solutions, namely,

$$x = 3 - 3k \qquad y = k \quad \text{for any } k$$

It is also interesting to note the numerators of the expressions for x and y. In the case of x, the numerator $ed - bf$ is precisely the determinant of the matrix obtained from **A** by substituting the column vector **B** for the first column of **A**. Similarly, the numerator of the expression for y is the determinant of the matrix obtained from **A** by substituting **B** for the second column of **A**. For example,

$$3x - 7y = 6$$
$$2x + 6y = 11$$

$$\mathbf{A} = \begin{pmatrix} 3 & -7 \\ 2 & 6 \end{pmatrix}$$

$$\det \mathbf{A} = (3)(6) - (2)(-7) = 18 + 14 = 32$$

$$x = \frac{\det \begin{pmatrix} 6 & -7 \\ 11 & 6 \end{pmatrix}}{32} = \frac{36 - (-7)(11)}{32} = \frac{113}{32}$$

$$y = \frac{\det \begin{pmatrix} 3 & 6 \\ 2 & 11 \end{pmatrix}}{32} = \frac{33 - 12}{32} = \frac{21}{32}$$

Another interesting aspect of the determinant of the 2×2 matrix is that it determines whether or not $\mathbf{A}^{-1}$ exists. Remember, given

$$\mathbf{A} = \begin{pmatrix} a & b \\ c & d \end{pmatrix}$$

to find $\mathbf{A}^{-1}$ we must solve

$$\begin{pmatrix} a & b \\ c & d \end{pmatrix}\begin{pmatrix} x \\ y \end{pmatrix} = \begin{pmatrix} 1 \\ 0 \end{pmatrix} \quad \text{and} \quad \begin{pmatrix} a & b \\ c & d \end{pmatrix}\begin{pmatrix} z \\ w \end{pmatrix} = \begin{pmatrix} 0 \\ 1 \end{pmatrix}$$

In order that unique solutions to these sets exist, we must have $\det \mathbf{A} \neq 0$. Thus, we have a very simple rule for determining whether or not a 2×2 matrix is invertible. If $\det \mathbf{A} = 0$, the matrix has no inverse. If $\det \mathbf{A} \neq 0$, the matrix has a unique inverse. In fact, solving

$$\begin{pmatrix} a & b \\ c & d \end{pmatrix}\begin{pmatrix} x \\ y \end{pmatrix} = \begin{pmatrix} 1 \\ 0 \end{pmatrix}$$

we obtain

$$x = \frac{\det \begin{pmatrix} 1 & b \\ 0 & d \end{pmatrix}}{\det \mathbf{A}} = \frac{d}{\det \mathbf{A}}$$

$$y = \frac{\det \begin{pmatrix} a & 1 \\ c & 0 \end{pmatrix}}{\det \mathbf{A}} = \frac{-c}{\det \mathbf{A}}$$

Similarly,

$$z = \frac{-b}{\det \mathbf{A}}$$

$$w = \frac{a}{\det \mathbf{A}}$$

You may recall that these equations arose by assuming $\mathbf{A}^{-1}$ of the form

$$\mathbf{A}^{-1} = \begin{pmatrix} x & z \\ y & w \end{pmatrix}$$

Hence,

$$\mathbf{A}^{-1} = \frac{1}{\det \mathbf{A}} \begin{pmatrix} d & -b \\ -c & a \end{pmatrix}$$

Here is one example:

$$\mathbf{A} = \begin{pmatrix} 3 & 2 \\ -1 & 4 \end{pmatrix}$$

$$\det \mathbf{A} = 12 + 2 = 14$$

and

$$\mathbf{A}^{-1} = \frac{1}{14} \begin{pmatrix} 4 & -2 \\ 1 & 3 \end{pmatrix}$$

EXERCISES

1. (a) Use determinants to solve for x and y: $4x - y = 10, 2x + 3y = 12$.
   (b) Use determinants to solve for x and y: $4x + 3y = -4$, $x - 2y = 10$.
2. Solve for x and y: $x^2 + 4y^2 = 8$, $2x^2 - y^2 = 7$.
3. Find the intersection of the two lines $y = 3x - 7$ and $y = x - 6$ by using determinants.

4. Solve for x:

$$\det\begin{pmatrix} 2x & -x \\ x & 3 \end{pmatrix} = 0$$

5. Given the matrices

$$\mathbf{A} = \begin{pmatrix} 4 & 3 \\ 1 & -2 \end{pmatrix} \quad \mathbf{B} = \begin{pmatrix} 5 & 2 \\ -4 & -1 \end{pmatrix} \quad \mathbf{C} = \begin{pmatrix} 6 & -4 \\ 0 & 1 \end{pmatrix}$$

find $\mathbf{A}^{-1}$, $\mathbf{B}^{-1}$ and $\mathbf{C}^{-1}$ by use of determinants.

6. Solve for x and y:

(a) $2/x + 3/y = 6$

$1/x - 1/y = 4$

(b) $2/x + 3/y = 6$

$1/x - 1/y = 3$

7. Show that

$$\det\begin{pmatrix} a & b \\ c & d \end{pmatrix} + \det\begin{pmatrix} e & b \\ f & d \end{pmatrix} = \det\begin{pmatrix} a+e & b \\ c+f & d \end{pmatrix}$$

and check that it works for

$$\det\begin{pmatrix} 1 & -1 \\ 2 & 3 \end{pmatrix} + \det\begin{pmatrix} -1 & -1 \\ 2 & 3 \end{pmatrix} = \det\begin{pmatrix} 0 & -1 \\ 4 & 3 \end{pmatrix}$$

8.2 Higher-order Determinants

It would be nice to extend the notion of the determinant to larger square matrices. We shall do so by defining the determinant to have certain properties. In particular, we shall describe what happens to the determinant of a matrix when it undergoes elementary row operations. However, let us first note the following:

$$\det\begin{pmatrix} a & b \\ c & d \end{pmatrix} = ad - bc \quad \text{and} \quad \det\begin{pmatrix} a & c \\ b & d \end{pmatrix} = ad - bc$$

That is, if we interchange the rows and columns of a matrix, the determinant is unchanged. What this means in practice is that any property of row operations would also be true if the corresponding operation were performed on the columns.

Take a simple row operation, multiplying one row by k:

$$\det\begin{pmatrix} ka & kb \\ c & d \end{pmatrix} = kad - kbc = k(ad - bc) = k \det\begin{pmatrix} a & b \\ c & d \end{pmatrix}$$

That is, multiplying one row by k multiplies the determinant by k. This also means that multiplying a column by k multiplies the determinant by k. Thus,

$$\det\begin{pmatrix} 2 & 4 & 6 \\ -1 & 8 & 7 \\ 2 & 0 & 8 \end{pmatrix} = 2\det\begin{pmatrix} 1 & 2 & 3 \\ -1 & 8 & 7 \\ 2 & 0 & 8 \end{pmatrix} = 4\det\begin{pmatrix} 1 & 1 & 3 \\ -1 & 4 & 7 \\ 2 & 0 & 8 \end{pmatrix}$$

where the first equality is obtained by factoring 2 from the first row and the second equality is obtained by factoring 2 from the second column. Let us take this to be our first property.

Property 1 If $\mathbf{A}$ is an $n \times n$ matrix and $\mathbf{B}$ is obtained from $\mathbf{A}$ by multiplying any row (or column) of $\mathbf{A}$ by a number k, then $\det \mathbf{B} = k \det \mathbf{A}$.

This leads automatically to the next property.

Property 2 If $\mathbf{A}$ has an entire row (or column) of zeros, then $\det \mathbf{A} = 0$.

Property 2 must be true since a matrix with a row of zeros may be thought of as obtained from another matrix by multiplying one row by zero and hence multiplying the determinant by zero.

Consider now the second row operation, interchanging two rows. If

$$\mathbf{A} = \begin{pmatrix} a & b \\ c & d \end{pmatrix} \quad \text{and} \quad \mathbf{B} = \begin{pmatrix} c & d \\ a & b \end{pmatrix}$$

then

$$\det \mathbf{A} = ad - bc \quad \text{and} \quad \det \mathbf{B} = bc - ad = -(ad - bc)$$

Thus

$$\det \mathbf{A} = -\det \mathbf{B}$$

and we define our next property.

Property 3 Interchanging two rows (or columns) of a matrix changes the sign of its determinant.

Consider next the third row operation, adding k times one row to another. That is,

$$\mathbf{A} = \begin{pmatrix} a & b \\ c & d \end{pmatrix} \quad \text{and} \quad \mathbf{B} = \begin{pmatrix} a + kc & b + kd \\ c & d \end{pmatrix}$$

Here,

$$\begin{aligned} \det \mathbf{B} &= (a + kc)d - (b + kd)c \\ &= ad + kcd - bc - kcd \\ &= ad - bc = \det A \end{aligned}$$

Thus we assume the following property.

Property 4 Adding any multiple of one row (or column) to another row (or column) leaves the determinant of the matrix unchanged.

This leads one automatically to Property 5.

Property 5 If two rows of a matrix are identical or if one row is a constant multiple of another row, then the determinant of the matrix is zero. (The same is true, of course, for columns.)

Property 5 is easy to prove since subtracting a multiple of one row from the other will leave a row of all zeros which we know by Property 2 gives a zero determinant.

We can now use row operations to attempt to diagonalize the matrix. As we know, sometimes this cannot be done because a row of all zeros will appear. In such cases, $\det A = 0$. If a row of zeros does not occur we will eventually, by total elimination, reduce **A** to **I**. In the 2×2 case

$$\det \mathrm{I} = \det \begin{pmatrix} 1 & 0 \\ 0 & 1 \end{pmatrix} = 1$$

and we shall assume this as the next property.

Property 6 $\det \mathbf{I} = 1$.

Please notice that Properties 1, 3, 4 and 6 are assumed and the other two are consequences. So far then we have simply assumed that four properties of 2×2 determinants should be properties of all $n \times n$ determinants and that these are sufficient to enable us to compute any determinant. For example, let

$$x = \det\begin{pmatrix} 1 & -1 & 2 \\ 3 & -1 & 1 \\ 1 & 2 & 4 \end{pmatrix}$$

Taking Column 2 + Column 1 and Column 3 − 2 × Column 1 leaves the determinant unchanged; so

$$x = \det\begin{pmatrix} 1 & -1 & 2 \\ 0 & 2 & -5 \\ 0 & 3 & 2 \end{pmatrix}$$

Taking Column 2 + Column 1 and Column 3 − 2 × Column 1 leaves the determinant unchanged; so

$$x = \det\begin{pmatrix} 1 & 0 & 0 \\ 0 & 2 & -5 \\ 0 & 3 & 2 \end{pmatrix}$$

Now, factor a 2 from Row 2:

$$x = 2\det\begin{pmatrix} 1 & 0 & 0 \\ 0 & 1 & -2.5 \\ 0 & 3 & 2 \end{pmatrix}$$

Taking Row 3 − 3 × Row 2 yields

$$x = 2\det\begin{pmatrix} 1 & 0 & 0 \\ 0 & 1 & -2.5 \\ 0 & 0 & 9.5 \end{pmatrix}$$

Factor 9.5 from Row 3:

$$x = 19\det\begin{pmatrix} 1 & 0 & 0 \\ 0 & 1 & -2.5 \\ 0 & 0 & 1 \end{pmatrix}$$

Take Row 2 + 2.5 × Row 3:

$$x = 19\det\begin{pmatrix} 1 & 0 & 0 \\ 0 & 1 & 0 \\ 0 & 0 & 1 \end{pmatrix} = 19$$

Some shortcuts should be noted. If we reach the form

$$\det \mathbf{A} = \det \begin{pmatrix} k_1 & 0 & 0 & \ldots & 0 \\ 0 & k_2 & 0 & \ldots & 0 \\ 0 & 0 & k_3 & \ldots & 0 \\ \vdots & \vdots & \vdots & & \vdots \\ 0 & 0 & 0 & \ldots & k_n \end{pmatrix}$$

then $\det \mathbf{A} = k_1 k_2 \ldots k_n$, since we simply factor k_1 from the first row, k_2 from the second and so on until reaching **I**. Therefore, the determinant of a diagonalized matrix may be found by inspection.

For an even easier method, if we reach

$$\det \mathbf{A} = \det \begin{pmatrix} a_1 & a_2 & a_3 & \ldots & a_n \\ 0 & b_2 & b_3 & \ldots & b_n \\ 0 & 0 & c_3 & \ldots & c_n \\ \vdots & \vdots & \vdots & & \vdots \\ 0 & 0 & 0 & \ldots & f_n \end{pmatrix}$$

then $\det \mathbf{A} = a_1 b_2 c_3 \ldots f_n$, that is, simply the product of the terms on the diagonal. This is so since if none of the terms on the diagonal is zero, we could have completed the diagonalization without changing any of them by simple row operations. If any is zero, then row operations will eventually yield a row of all zeros and $\det \mathbf{A} = 0$ as would be the product $a_1 b_2 c_3, \ldots, f_n$. Therefore, we again have the nice situation where triangularization will suffice. By the way, we could also triangularize to have all zeros in the upper right instead of lower left, since for determinants the roles of row and columns may be interchanged.

Try to follow the operations in the following example:

$$\det \begin{pmatrix} 1 & -1 & 2 & 3 \\ 1 & -1 & 1 & 6 \\ 2 & -3 & 1 & 4 \\ -1 & 1 & 2 & 0 \end{pmatrix} = \det \begin{pmatrix} 1 & -1 & 2 & 3 \\ 0 & 0 & -1 & 3 \\ 0 & -1 & -3 & -2 \\ 0 & 0 & 4 & 3 \end{pmatrix}$$

$$= -\det \begin{pmatrix} 1 & -1 & 2 & 3 \\ 0 & -1 & -3 & -2 \\ 0 & 0 & -1 & 3 \\ 0 & 0 & 4 & 3 \end{pmatrix}$$

$$= -(-1)\det \begin{pmatrix} 1 & -1 & 2 & 3 \\ 0 & 1 & 3 & 2 \\ 0 & 0 & -1 & 3 \\ 0 & 0 & 4 & 3 \end{pmatrix}$$

$$= \det \begin{pmatrix} 1 & -1 & 2 & 3 \\ 0 & 1 & 3 & 2 \\ 0 & 0 & -1 & 3 \\ 0 & 0 & 0 & 15 \end{pmatrix}$$

$$= (1)(1)(-1)(15) = -15$$

Let us review the steps taken. First, we used row operations to clear the first column. Next, we interchanged the first and second rows. This changed the sign of the determinant. In the third step, we factored -1 from the second row. Finally, we cleared the third column and having triangularized the matrix, we evaluated the determinant by taking the product of the terms on the diagonal.

EXERCISES

Evaluate the determinants of the following matrices:

1. $\begin{pmatrix} -1 & 2 & 1 \\ 0 & 3 & 7 \\ 0 & 0 & 6 \end{pmatrix}$

2. $\begin{pmatrix} 1 & 1/2 & 7 \\ 2 & 1 & 14 \\ 11 & 3 & 6 \end{pmatrix}$

3. $\begin{pmatrix} 2 & 0 & 0 \\ -3 & 6 & 0 \\ 11 & 14 & 4 \end{pmatrix}$

4. $\begin{pmatrix} 1 & -2 & 1 \\ 6 & 3 & 6 \\ 9 & 11 & 9 \end{pmatrix}$

5. $\begin{pmatrix} 1 & 2 & -1 \\ 1 & -1 & 3 \\ 0 & 2 & 1 \end{pmatrix}$

6. $\begin{pmatrix} 2 & 0 & -3 \\ 1 & 1 & 2 \\ 0 & -3 & 1 \end{pmatrix}$

7. $\begin{pmatrix} 2 & 0 & -3 & 1 \\ 1 & 1 & 2 & -2 \\ 0 & -3 & 1 & 4 \\ 1 & 0 & 2 & 3 \end{pmatrix}$

8. $\begin{pmatrix} 1 & 1 & 0 & 3 \\ -2 & -1 & 3 & -4 \\ 2 & 2 & -2 & 3 \\ -1 & -2 & 1 & -2 \end{pmatrix}$

Solve for x in the following:

9. $\det\begin{pmatrix} 1 & -1 & 2 \\ 3 & 0 & 2 \\ -1 & 4 & 3 \end{pmatrix} = \det\begin{pmatrix} -1 & -1 & 2 \\ 3 & 0 & 2 \\ x & 4 & 3 \end{pmatrix}$

10. $\det\begin{pmatrix} 1 & 6 & 4 \\ 2 & 3 & 0 \\ -1 & 1 & 2 \end{pmatrix} = \det\begin{pmatrix} 1 & 7 & 4 \\ 2 & 5 & 0 \\ -1 & x & 2 \end{pmatrix}$

11. $\det\begin{pmatrix} 1 & 2 & 3 \\ 0 & x & 4 \\ 0 & 0 & x \end{pmatrix} = 9$

12. $\det\begin{pmatrix} x & 2 & 3 \\ 0 & 1 & -1 \\ 0 & 4 & x \end{pmatrix} = -3$

13. $\det\begin{pmatrix} 1 & 2 & x \\ 2 & -1 & 0 \\ 2x+9 & 6 & -9 \end{pmatrix} = 0$

8.3 Further Properties of the Determinant

If $\mathbf{A}$ is a square matrix for which no inverse exists, we shall say that $\mathbf{A}$ is *singular*. If $\mathbf{A}^{-1}$ exists, we shall say that $\mathbf{A}$ is nonsingular. If $\mathbf{A}$ is nonsingular, then we may find row operations which reduce $\mathbf{A}$ to $\mathbf{I}$ and hence $\det \mathbf{A}$ will be some multiple of $\det \mathbf{I}$ which is 1. Hence, if $\mathbf{A}$ is nonsingular, $\det \mathbf{A} \neq 0$. If $\mathbf{A}$ is singular, then a sequence of row operations will reduce $\mathbf{A}$ to a matrix with a row of all zeros and which, therefore, has determinant zero. Hence, if $\mathbf{A}$ is singular, then $\det \mathbf{A} = 0$. This gives us a very nice result.

THEOREM 1 If $\mathbf{A}$ is an $n \times n$ matrix, then $\mathbf{A}^{-1}$ exists if and only if $\det \mathbf{A} \neq 0$.

Furthermore, suppose that $\mathbf{A}$ and $\mathbf{B}$ are $n \times n$ matrices and let $\mathbf{C} = \mathbf{AB}$. If $\mathbf{A}^{-1}$ and $\mathbf{B}^{-1}$ both exist, then consider the product $\mathbf{B}^{-1}\mathbf{A}^{-1}\mathbf{C}$. We have,

$$\begin{aligned} \mathbf{B}^{-1}\mathbf{A}^{-1}\mathbf{C} &= \mathbf{B}^{-1}\mathbf{A}^{-1}(\mathbf{AB}) = \mathbf{B}^{-1}(\mathbf{A}^{-1}\mathbf{A})\mathbf{B} = \mathbf{B}^{-1}\mathbf{IB} \\ &= \mathbf{B}^{-1}\mathbf{B} = \mathbf{I} \end{aligned}$$

Hence, $\mathbf{B}^{-1}\mathbf{A}^{-1} = \mathbf{C}^{-1}$ and we have the following corollary.

Corollary 1 If **A** and **B** are nonsingular, then **AB** is nonsingular.

On the other hand, suppose **A** is singular. Then we shall show that the product **AB** is also singular. Suppose $\mathbf{C} = \mathbf{AB}$ and $\mathbf{C}^{-1}$ exists; then

$$(\mathbf{AB})\mathbf{C}^{-1} = \mathbf{CC}^{-1} = \mathbf{I}$$

that is,

$$\mathbf{A}(\mathbf{BC}^{-1}) = \mathbf{I}$$

But this means that $\mathbf{BC}^{-1}$ is an inverse for **A**, contradicting the assumption that **A** is singular. Similarly, if we assume **B** to be singular we can show that **AB** will be singular. This gives us the following.

THEOREM 2 If **A** and **B** are $n \times n$ matrices and $\det \mathbf{AB} = 0$, then $\det \mathbf{A} = 0$ or $\det \mathbf{B} = 0$ (or both).

Theorem 2 is analogous to the case of numbers for which $ab = 0$ means that $a = 0$ or $b = 0$, or both. In fact, we can make even stronger statements. First, let us define "elementary matrix." If **I** is an identity matrix and **E** is a matrix obtained from **I** by performing an elementary row operation, then **E** is called an *elementary* matrix. For 3×3 matrices

$$\mathbf{E}_1 = \begin{pmatrix} 0 & 1 & 0 \\ 1 & 0 & 0 \\ 0 & 0 & 1 \end{pmatrix} \quad \text{obtained by interchanging Rows 1 and 2}$$

$$\mathbf{E}_2 = \begin{pmatrix} 1 & 0 & 0 \\ 0 & 1 & 0 \\ 0 & 0 & -2 \end{pmatrix} \quad \text{obtained by multiplying Row 3 by } -2$$

and

$$\mathbf{E}_3 = \begin{pmatrix} 1 & 0 & 0 \\ 0 & 1 & 0 \\ -4 & 0 & 1 \end{pmatrix} \quad \text{obtained by adding } -4 \text{ times Row 1 to Row 3}$$

are all examples of elementary matrices. Now, $\det \mathbf{E}_1 = -1$, $\det \mathbf{E}_2 = -2$ and $\det \mathbf{E}_3 = 1$.

Suppose **E** is obtained from **I** by performing a particular elementary row operation on **I**. Suppose **B** is the matrix obtained from **A** by performing the same operation on **A**. Then, it can be shown that $\mathbf{B} = \mathbf{EA}$.

If the operation is interchange of two rows, then $\det \mathbf{E} = -1$ and $\det \mathbf{B} = -\det \mathbf{A}$; hence $\det \mathbf{B} = \det \mathbf{E} \cdot \det \mathbf{A}$. Similarly, if the operation is multiplication of a row by k, then $\det \mathbf{E} = k$, $\det \mathbf{B} = k \cdot \det \mathbf{A}$ and hence $\det \mathbf{B} = \det \mathbf{E} \cdot \det \mathbf{A}$. And, of course, if the operation is adding a multiple of one row to another, $\det \mathbf{E} = 1$, $\det \mathbf{B} = \det \mathbf{A}$ and hence $\det \mathbf{B} = \det \mathbf{E} \cdot \det \mathbf{A}$. Thus, we may conclude that if $\mathbf{E}$ is an elementary matrix, $\det \mathbf{EA} = \det \mathbf{E} \cdot \det \mathbf{A}$.

It is also easy to see that the inverse of any elementary matrix is also an elementary matrix. That is, to undo the interchange of two rows, interchange them again. To undo multiplication by k, multiply by $1/k$ and so forth.

Now if $\mathbf{A}^{-1}$ exists, then there exists a sequence of row operations which reduces $\mathbf{A}$ to $\mathbf{I}$. So there is a sequence of elementary matrices $\mathbf{E}_1, \mathbf{E}_2 \ldots \mathbf{E}_n$ such that $\mathbf{E}_1\mathbf{E}_2 \ldots \mathbf{E}_n\mathbf{A} = \mathbf{I}$. Hence,

$$\mathbf{A} = \mathbf{E}_n^{-1}\mathbf{E}_{n-1}^{-1} \ldots \mathbf{E}_2^{-1}\mathbf{E}_1^{-1} = \mathbf{F}_1\mathbf{F}_2 \ldots \mathbf{F}_n$$

where

$$\mathbf{F}_1 = \mathbf{E}_n^{-1}, \mathbf{F}_2 = \mathbf{E}_{n-1}^{-1}, \ldots, \mathbf{F}_n = \mathbf{E}_1^{-1}$$

and

$$\det \mathbf{A} = \det \mathbf{F}_1 \cdot \det \mathbf{F}_2 \cdot \ldots \cdot \det \mathbf{F}_n$$

If $\mathbf{B}^{-1}$ exists, then similarly

$$\mathbf{B} = \mathbf{G}_1\mathbf{G}_2 \ldots \mathbf{G}_m \quad \text{and} \quad \det \mathbf{B} = \det \mathbf{G}_1 \cdot \det \mathbf{G}_2 \cdot \ldots \cdot \det \mathbf{G}_m$$

Finally,

$$\begin{aligned} \det \mathbf{AB} &= \det \mathbf{F}_1\mathbf{F}_2 \ldots \mathbf{F}_n\mathbf{G}_1\mathbf{G}_2, \ldots, \mathbf{G}_m \\ &= \det \mathbf{F}_1 \cdot \ldots \cdot \det \mathbf{F}_n \cdot \det \mathbf{G}_1 \cdot \ldots \cdot \det \mathbf{G}_m \\ &= \det \mathbf{A} \cdot \det \mathbf{B} \end{aligned}$$

Hence for nonsingular matrices,

$$\det \mathbf{AB} = \det \mathbf{A} \cdot \det \mathbf{B}$$

For singular matrices we have already seen this to be true, since both sides must be zero. Therefore, we have the following theorem.

THEOREM 3 If $\mathbf{A}$ and $\mathbf{B}$ are $n \times n$ matrices, $\det \mathbf{AB} = \det \mathbf{A} \cdot \det \mathbf{B}$.

We shall now state one last result without proof. Recall that for 2×2 determinants, we showed that the solutions to

$$\begin{aligned} ax + by &= e \\ cx + dy &= f \end{aligned}$$

are

$$x = \frac{\det \begin{pmatrix} e & b \\ f & d \end{pmatrix}}{\det \begin{pmatrix} a & b \\ c & d \end{pmatrix}} \qquad y = \frac{\det \begin{pmatrix} a & e \\ c & f \end{pmatrix}}{\det \begin{pmatrix} a & b \\ c & d \end{pmatrix}}$$

The generalization of this follows.

Cramer's Rule If $\mathbf{AX} = \mathbf{B}$ is a system of n equations in n unknowns,

$$\mathbf{X} = \begin{pmatrix} x_1 \\ x_2 \\ \vdots \\ x_n \end{pmatrix}$$

and $\det \mathbf{A} = D \neq 0$. Let $\mathbf{A}_i$ be the matrix obtained from $\mathbf{A}$ by replacing the ith column of $\mathbf{A}$ by $\mathbf{B}$. Let $D_i = \det \mathbf{A}_i$. Then $x_i = D_i/D$.

Take a 3×3 example:

$$\begin{aligned} 2x - y + z &= 3 \\ 2x + z &= 4 \\ x + 2y - z &= 6 \end{aligned}$$

$$D = \det \begin{pmatrix} 2 & -1 & 1 \\ 2 & 0 & 1 \\ 1 & 2 & -1 \end{pmatrix} = -3$$

$$D_1 = \det \begin{pmatrix} 3 & -1 & 1 \\ 4 & 0 & 1 \\ 6 & 2 & -1 \end{pmatrix} = -8 \qquad x = \frac{-8}{-3} = \frac{8}{3}$$

$$D_2 = \det \begin{pmatrix} 2 & 3 & 1 \\ 2 & 4 & 1 \\ 1 & 6 & -1 \end{pmatrix} = -3 \qquad y = \frac{-3}{-3} = 1$$

$$D_3 = \det \begin{pmatrix} 2 & -1 & 3 \\ 2 & 0 & 4 \\ 1 & 2 & 6 \end{pmatrix} = 4 \qquad z = \frac{4}{-3} = \frac{-4}{3}$$

EXERCISES

1. For which of the following does the inverse matrix exist:

$$\mathbf{A} = \begin{pmatrix} 2 & -1 & 0 \\ 2 & 1 & 1 \\ 3 & -1 & 2 \end{pmatrix} \qquad \mathbf{B} = \begin{pmatrix} -1 & -1 & -1 \\ 1 & 1 & 1 \\ 0 & 3 & 5 \end{pmatrix}$$

$$\mathbf{C} = \begin{pmatrix} 0 & 3 & 0 \\ 1 & 8 & -2 \\ 4 & -1 & 6 \end{pmatrix} \qquad \mathbf{D} = \begin{pmatrix} b & a & a & 1 \\ a & b & a & 1 \\ a & a & b & 1 \\ a & a & a & 1 \end{pmatrix}$$

2. For $\mathbf{A}$ and $\mathbf{C}$ in Exercise 1, show that $\det \mathbf{AC} = \det \mathbf{A} \cdot \det \mathbf{C}$.

3. Use determinants to solve

$$\begin{aligned} x + 2y + 2z &= 4 \\ y + 4z &= 3 \\ 3x - 2y + z &= -2 \end{aligned}$$

4. Use determinants to solve

$$\begin{aligned} 2x + y &= 1 \\ x - z &= 1 \\ 2y + 3z &= -3 \end{aligned}$$

5. Use determinants to solve

$$\begin{aligned} 2x - y + 2z &= 4 \\ x + 2y + 3z &= 1 \\ x - y &= 1 \end{aligned}$$

6. For $\mathbf{A}$ in Exercise 1 and $\mathbf{E}_1$, $\mathbf{E}_2$ and $\mathbf{E}_3$ given in the text, show that $\mathbf{E}_1\mathbf{A}$ gives $\mathbf{A}$ with Rows 1 and 2 interchanged. Show similarly for $\mathbf{E}_2\mathbf{A}$ and $\mathbf{E}_3\mathbf{A}$. What happens when you form $\mathbf{AE}_1$, $\mathbf{AE}_2$ and $\mathbf{AE}_3$?

7. Show that $\mathbf{E}_1^{-1} = \mathbf{E}_1$, for $\mathbf{E}_1$ given in the text. What are $\mathbf{E}_2^{-1}$ and $\mathbf{E}_3^{-1}$?

8. For $\mathbf{A}$ and $\mathbf{C}$ in Exercise 1, show that $(\mathbf{AC})^{-1} = \mathbf{C}^{-1}\mathbf{A}^{-1}$.

9. For $\mathbf{A}$ in Exercise 1, show that $\det(\mathbf{A}^2) = [\det \mathbf{A}]^2$.

10. For $\mathbf{A}$ and $\mathbf{C}$ in Exercise 1, show that $\det(\mathbf{A} + \mathbf{C}) \neq \det \mathbf{A} + \det \mathbf{C}$.

11. Construct two 3×3 matrices such that $\det \mathbf{A} = 0$, $\det \mathbf{B} = 0$ and $\det(\mathbf{A} + \mathbf{B}) \neq 0$.

12. Solve for x, $\det(x\mathbf{A} + \mathbf{B}) = 0$, where

$$\mathbf{A} = \begin{pmatrix} -1 & 2 \\ 1 & 3 \end{pmatrix} \quad \text{and} \quad \mathbf{B} = \begin{pmatrix} 1 & -1 \\ 3 & 0 \end{pmatrix}$$

13. Show that $\det(\mathbf{A}^{-1}) = [\det \mathbf{A}]^{-1}$.

14. Show that $\det(k\mathbf{A}) = k^n \det \mathbf{A}$, where $\mathbf{A}$ is $n \times n$.

15. Markov chains. To find the fixed point for a regular Markov chain, it is necessary to solve

$$\mathbf{wP} = \mathbf{w}$$

This may be rewritten as

$$\mathbf{wP} - \mathbf{wI} = \mathbf{0}$$
$$\mathbf{w}(\mathbf{P} - \mathbf{I}) = \mathbf{0}$$

If $(\mathbf{P} - \mathbf{I})^{-1}$ exists, then $\mathbf{w} = \mathbf{0}$ is the only solution. Show by determinants that for any two-state Markov process, $(\mathbf{P} - \mathbf{I})^{-1}$ does not exist.

16. Repeat Exercise 15 for a three-state process. Can you find a general proof?

17. Suppose $\mathbf{A} = \mathbf{A}^{-1}$; show that $\det \mathbf{A} = \pm 1$. Give an example of such a matrix.

18. Find $\det \begin{pmatrix} 1 & b \\ b & 1 \end{pmatrix}$, $\det \begin{pmatrix} 1 & b & b \\ b & 1 & b \\ b & b & 1 \end{pmatrix}$. Can you find a general formula for

$$\det \begin{pmatrix} 1 & b & \dots & b \\ b & 1 & \dots & b \\ b & b & \dots & b \\ \vdots & \vdots & & \vdots \\ b & b & \dots b & 1 \end{pmatrix}$$

Appendix 1

PROOF OF CHEBYSCHEV'S INEQUALITY

Let $R = \{x_1, x_2, \ldots, x_n\}$ be the range of X and suppose that $\{x_1, x_2, \ldots, x_m\}$ lie outside the interval

$$\mu - k\sigma \leq x \leq \mu + k\sigma$$

Now

$$P[\mu - k\sigma \leq X \leq \mu + k\sigma] = 1 - [p_1 + p_2 + \cdots + p_m]$$

For $j = 1, 2, \ldots, m$

$$(x_j - \mu)^2 \geq k^2\sigma^2$$

Therefore,

$$\left(\frac{x_j - \mu}{k\sigma}\right)^2 \geq 1$$

and hence

$$\begin{aligned} p_1 + p_2 + \cdots + p_m &\leq \left(\frac{x_1 - \mu}{k\sigma}\right)^2 p_1 + \cdots + \left(\frac{x_m - \mu}{k\sigma}\right)^2 p_m \\ &= \frac{1}{k^2\sigma^2}\,[(x_1 - \mu)^2 p_1 + \cdots + (x_m - \mu)^2 p_m)] \end{aligned}$$

Since each term $(x_j - \mu)^2 p_j / k^2\sigma^2$ is positive, we may now add the remaining terms, which yields

$$\begin{aligned} p_1 + p_2 + \cdots + p_m \leq \frac{1}{k^2\sigma^2}\,[(x_1 - \mu)^2 p_1 &+ (x_2 - \mu)^2 p_2 \\ &+ \cdots + (x_m - \mu)^2 p_m + \cdots + (x_n - \mu)^2 p_n] \end{aligned}$$

The final term in square brackets we recognize to be by definition σ^2. Hence,

$$p_1 + p_2 + \cdots + p_m \leq \frac{1}{k^2\sigma^2}\,\sigma^2 = \frac{1}{k^2}$$

Thus

$$P[\mu - k\sigma \leq X \leq \mu + k\sigma] \geq 1 - \frac{1}{k^2}$$

Appendix 2

RELATIONS AND FUNCTIONS

Suppose $A = \{x : 0 \leq x \leq 2\}$ and $B = \{y : 0 \leq y \leq 3\}$. There are various rules by which we may associate elements of A and elements of B. Any specific rule is called a *relation*. For example, the rule $y \leq x$ associates with every $x \in A$ all those elements of B which are less than or equal to x. In this relation there are many y-values associated with each x-value. The set of x-values for which the rule is defined consists of all of the set A. The set of y-values for which the rule is defined consists of a subset of B, namely, $\{y : 0 \leq y \leq 2\}$.

Consider now the rule $y = x^2$. This rule makes sense for all $x \in A$, but in order that the resulting y be an element of B, we must restrict our attention to $D = \{x : 0 \leq x \leq \sqrt{3}\}$. On the other hand, the set of possible y-values is all of B. However, the characteristic of greatest interest in this relation is that to each value of x, for which the relation is defined, there is associated exactly one y-value. Such a relation is called a *function*. The set of x-values for which the function is defined is called the *domain* of the function. The set of y-values for which the function is defined is called the *range* of the function. As one allows x to vary, y varies in an exactly predictable fashion. This is so because for every chosen x, precisely one y-value is defined. Therefore, we call x the *independent* variable, y the *dependent* variable, and we say that y is a function of x.

We usually use the following notational conventions. The function is called f (or g, or h, . . .); the set D of possible values of the independent variable is called the *domain* of f. The value of y associated with a given x is denoted by $f(x)$ (read "f of x") and we write the rule as $y = f(x)$. Finally, the set of possible y-values is called the *range* of f. Mathematically, we could describe the range as

$$R = \{y : y = f(x), x \in D\}$$

For example, if the rule is $y = x^{1/2}$ and if we deal only with real numbers, then the largest set that D could be is $D = \{x : x \geq 0\}$, since we cannot take the square root of a negative number. In $f(x)$ notation, we would say that $f(x) = x^{1/2}$. In this case $R = \{y : y \geq 0\}$.

From these examples it would appear that D and f determine R, and this is correct. However, it is also possible that R may determine D to some extent, as we can see from the earlier example, $y = x^2$, where $0 \leq y \leq 3$ forced $0 \leq x \leq \sqrt{3}$.

The principal difference between the independent and dependent variable is as follows: For every x there is associated only one y but for several different x's, $f(x)$ may be the same. For example, if $y = x^2$ and the domain contains positive and negative values, we have $f(-a) = (-a)^2 = a^2$ and $f(a) = a^2$ for every choice of a. That is, each y-value (except 0) is associated with two different x-values.

In this book, functions appear in a variety of guises. The first function actually identified as such is the random variable. For the random variable we use a markedly different notation from that given here. In addition, the domain of the random variable is not a set of numbers. It is instead a set of elements in a sample space. Prior to the random variable we encounter the probability measure. This is also a function but we do not refer to it as such in the text. This function has as its domain the space of subsets of the sample space. That is, the points in the domain are sets. We also encounter the objective function for a linear programming problem. This function has as its domain the space of ordered pairs or triples (or n-tuples), depending upon the number of variables. We even use the determinant function which is defined on the space of square matrices. Thus there is almost no limit to the kinds of mathematical spaces upon which functions may be defined.

Appendix 3

TABLES

Table 1

Binomial Probabilities*

n	k	.05	.10	.15	.20	.25	.30	.35	.40	.45	.50
1	0	.9500	.9000	.8500	.8000	.7500	.7000	.6500	.6000	.5500	.5000
	1	.0500	.1000	.1500	.2000	.2500	.3000	.3500	.4000	.4500	.5000
2	0	.9025	.8100	.7225	.6400	.5625	.4900	.4225	.3600	.3025	.2500
	1	.0950	.1800	.2550	.3200	.3750	.4200	.4550	.4800	.4950	.5000
	2	.0025	.0100	.0225	.0400	.0625	.0900	.1225	.1600	.2025	.2500
3	0	.8574	.7290	.6141	.5120	.4219	.3430	.2746	.2160	.1664	.1250
	1	.1354	.2430	.3251	.3840	.4219	.4410	.4436	.4320	.4084	.3750
	2	.0071	.0270	.0574	.0960	.1406	.1890	.2389	.2880	.3341	.3750
	3	.0001	.0010	.0034	.0080	.0156	.0270	.0429	.0640	.0911	.1250
4	0	.8145	.6561	.5220	.4096	.3164	.2401	.1785	.1296	.0915	.0625
	1	.1715	.2916	.3685	.4096	.4219	.4116	.3845	.3456	.2995	.2500
	2	.0135	.0486	.0975	.1536	.2109	.2646	.3105	.3456	.3675	.3750
	3	.0005	.0036	.0115	.0256	.0469	.0756	.1115	.1536	.2005	.2500
	4	.0000	.0001	.0005	.0016	.0039	.0081	.0150	.0256	.0410	.0625
5	0	.7738	.5905	.4437	.3277	.2373	.1681	.1160	.0778	.0503	.0312
	1	.2036	.3280	.3915	.4096	.3955	.3602	.3124	.2592	.2059	.1562
	2	.0214	.0729	.1382	.2048	.2637	.3087	.3364	.3456	.3369	.3125
	3	.0011	.0081	.0244	.0512	.0879	.1323	.1811	.2304	.2757	.3125
	4	.0000	.0004	.0022	.0064	.0146	.0284	.0488	.0768	.1128	.1562
	5	.0000	.0000	.0001	.0003	.0010	.0024	.0053	.0102	.0185	.0312
6	0	.7351	.5314	.3771	.2621	.1780	.1176	.0754	.0467	.0277	.0156
	1	.2321	.3543	.3993	.3932	.3560	.3025	.2437	.1866	.1359	.0938
	2	.0305	.0984	.1762	.2458	.2966	.3241	.3280	.3110	.2780	.2344
	3	.0021	.0146	.0415	.0819	.1318	.1852	.2355	.2765	.3032	.3125
	4	.0001	.0012	.0055	.0154	.0330	.0595	.0951	.1382	.1861	.2344
	5	.0000	.0001	.0004	.0015	.0044	.0102	.0205	.0369	.0609	.0938
	6	.0000	.0000	.0000	.0001	.0002	.0007	.0018	.0041	.0083	.0156

* Entries in this table are values of $\binom{n}{k}p^k(1-p)^{n-k}$ for the indicated values of n, k and p.

Table 1 (continued)

n	k	p .05	.10	.15	.20	.25	.30	.35	.40	.45	.50
7	0	.6983	.4783	.3206	.2097	.1335	.0824	.0490	.0280	.0152	.0078
	1	.2573	.3720	.3960	.3670	.3115	.2471	.1848	.1306	.0872	.0547
	2	.0406	.1240	.2097	.2753	.3115	.3177	.2985	.2613	.2140	.1641
	3	.0036	.0230	.0617	.1147	.1730	.2269	.2679	.2903	.2918	.2734
	4	.0002	.0026	.0109	.0287	.0577	.0972	.1442	.1935	.2388	.2734
	5	.0000	.0002	.0012	.0043	.0115	.0250	.0466	.0774	.1172	.1641
	6	.0000	.0000	.0001	.0004	.0013	.0036	.0084	.0172	.0320	.0547
	7	.0000	.0000	.0000	.0000	.0001	.0002	.0006	.0016	.0037	.0078
8	0	.6634	.4305	.2725	.1678	.1001	.0576	.0319	.0168	.0084	.0039
	1	.2793	.3826	.3847	.3355	.2670	.1977	.1373	.0896	.0548	.0312
	2	.0515	.1488	.2376	.2936	.3115	.2965	.2587	.2090	.1569	.1094
	3	.0054	.0331	.0839	.1468	.2076	.2541	.2786	.2787	.2568	.2188
	4	.0004	.0046	.0185	.0459	.0865	.1361	.1875	.2322	.2627	.2734
8	5	.0000	.0004	.0026	.0092	.0231	.0467	.0808	.1239	.1719	.2188
	6	.0000	.0000	.0002	.0011	.0038	.0100	.0217	.0413	.0703	.1094
	7	.0000	.0000	.0000	.0001	.0004	.0012	.0033	.0079	.0164	.0312
	8	.0000	.0000	.0000	.0000	.0000	.0001	.0002	.0007	.0017	.0039
9	0	.6302	.3874	.2316	.1342	.0751	.0404	.0207	.0101	.0046	.0020
	1	.2985	.3874	.3679	.3020	.2253	.1556	.1004	.0605	.0339	.0176
	2	.0629	.1722	.2597	.3020	.3003	.2668	.2162	.1612	.1110	.0703
	3	.0077	.0446	.1069	.1762	.2336	.2668	.2716	.2508	.2119	.1641
	4	.0006	.0074	.0283	.0661	.1168	.1715	.2194	.2508	.2600	.2461
	5	.0000	.0008	.0050	.0165	.0389	.0735	.1181	.1672	.2128	.2461
	6	.0000	.0001	.0006	.0028	.0087	.0210	.0424	.0743	.1160	.1641
	7	.0000	.0000	.0000	.0003	.0012	.0039	.0098	.0212	.0407	.0703
	8	.0000	.0000	.0000	.0000	.0001	.0004	.0013	.0035	.0083	.0176
	9	.0000	.0000	.0000	.0000	.0000	.0000	.0001	.0003	.0008	.0020
10	0	.5987	.3487	.1969	.1074	.0563	.0282	.0135	.0060	.0025	.0010
	1	.3151	.3874	.3474	.2684	.1877	.1211	.0725	.0403	.0207	.0098
	2	.0746	.1937	.2759	.3020	.2816	.2335	.1757	.1209	.0763	.0439
	3	.0105	.0574	.1298	.2013	.2503	.2668	.2522	.2150	.1665	.1172
	4	.0010	.0112	.0401	.0881	.1460	.2001	.2377	.2508	.2384	.2051

Table 1 (continued)

n	k	.05	.10	.15	.20	.25	.30	.35	.40	.45	.50
						p					
10	5	.0001	.0015	.0085	.0264	.0584	.1029	.1536	.2007	.2340	.2461
	6	.0000	.0001	.0012	.0055	.0162	.0368	.0689	.1115	.1596	.2051
	7	.0000	.0000	.0001	.0008	.0031	.0090	.0212	.0425	.0746	.1172
	8	.0000	.0000	.0000	.0001	.0004	.0014	.0043	.0106	.0229	.0439
	9	.0000	.0000	.0000	.0000	.0000	.0001	.0005	.0016	.0042	.0098
	10	.0000	.0000	.0000	.0000	.0000	.0000	.0000	.0001	.0003	.0010
11	0	.5688	.3138	.1673	.0859	.0422	.0198	.0088	.0036	.0014	.0005
	1	.3293	.3835	.3248	.2362	.1549	.0932	.0518	.0266	.0125	.0054
	2	.0867	.2131	.2866	.2953	.2581	.1998	.1395	.0887	.0513	.0269
	3	.0137	.0710	.1517	.2215	.2581	.2568	.2254	.1774	.1259	.0806
	4	.0014	.0158	.0536	.1107	.1721	.2201	.2428	.2365	.2060	.1611
	5	.0001	.0025	.0132	.0388	.0803	.1321	.1830	.2207	.2360	.2256
	6	.0000	.0003	.0023	.0097	.0268	.0566	.0985	.1471	.1931	.2256
	7	.0000	.0000	.0003	.0017	.0064	.0173	.0379	.0701	.1128	.1611
	8	.0000	.0000	.0000	.0002	.0011	.0037	.0102	.0234	.0462	.0806
	9	.0000	.0000	.0000	.0000	.0001	.0005	.0018	.0052	.0126	.0269
	10	.0000	.0000	.0000	.0000	.0000	.0000	.0002	.0007	.0021	.0054
	11	.0000	.0000	.0000	.0000	.0000	.0000	.0000	.0000	.0002	.0005
12	0	.5404	.2824	.1422	.0687	.0317	.0138	.0057	.0022	.0008	.0002
	1	.3413	.3766	.3012	.2062	.1267	.0712	.0368	.0174	.0075	.0029
	2	.0988	.2301	.2924	.2835	.2323	.1678	.1088	.0639	.0339	.0161
	3	.0173	.0852	.1720	.2362	.2581	.2397	.1954	.1419	.0923	.0537
	4	.0021	.0213	.0683	.1329	.1936	.2311	.2367	.2128	.1700	.1208
	5	.0002	.0038	.0193	.0532	.1032	.1585	.2039	.2270	.2225	.1934
	6	.0000	.0005	.0040	.0155	.0401	.0792	.1281	.1766	.2124	.2256
	7	.0000	.0000	.0006	.0033	.0115	.0291	.0591	.1009	.1489	.1934
	8	.0000	.0000	.0001	.0005	.0024	.0078	.0199	.0420	.0762	.1208
	9	.0000	.0000	.0000	.0001	.0004	.0015	.0048	.0125	.0277	.0537
	10	.0000	.0000	.0000	.0000	.0000	.0002	.0008	.0025	.0068	.0161
	11	.0000	.0000	.0000	.0000	.0000	.0000	.0001	.0003	.0010	.0029
	12	.0000	.0000	.0000	.0000	.0000	.0000	.0000	.0000	.0001	.0002
13	0	.5133	.2542	.1209	.0550	.0238	.0097	.0037	.0013	.0004	.0001
	1	.3512	.3672	.2774	.1787	.1029	.0540	.0259	.0113	.0045	.0016
	2	.1109	.2448	.2937	.2680	.2059	.1388	.0836	.0453	.0220	.0095
	3	.0214	.0997	.1900	.2457	.2517	.2181	.1651	.1107	.0660	.0349
	4	.0028	.0277	.0838	.1535	.2097	.2337	.2222	.1845	.1350	.0873

Table 1 (continued)

n	k	p .05	.10	.15	.20	.25	.30	.35	.40	.45	.50
13	5	.0003	.0055	.0266	.0691	.1258	.1803	.2154	.2214	.1989	.1571
	6	.0000	.0008	.0063	.0230	.0559	.1030	.1546	.1968	.2169	.2095
	7	.0000	.0001	.0011	.0058	.0186	.0442	.0833	.1312	.1775	.2095
	8	.0000	.0000	.0001	.0011	.0047	.0142	.0336	.0656	.1089	.1571
	9	.0000	.0000	.0000	.0001	.0009	.0034	.0101	.0243	.0495	.0873
	10	.0000	.0000	.0000	.0000	.0001	.0006	.0022	.0065	.0162	.0349
	11	.0000	.0000	.0000	.0000	.0000	.0001	.0003	.0012	.0036	.0095
	12	.0000	.0000	.0000	.0000	.0000	.0000	.0000	.0001	.0005	.0016
	13	.0000	.0000	.0000	.0000	.0000	.0000	.0000	.0000	.0000	.0001
14	0	.4877	.2288	.1028	.0440	.0178	.0068	.0024	.0008	.0002	.0001
	1	.3593	.3559	.2539	.1539	.0832	.0407	.0181	.0073	.0027	.0009
	2	.1229	.2570	.2912	.2501	.1802	.1134	.0634	.0317	.0141	.0056
	3	.0259	.1142	.2056	.2501	.2402	.1943	.1366	.0845	.0462	.0222
	4	.0037	.0349	.0998	.1720	.2202	.2290	.2022	.1549	.1040	.0611
	5	.0004	.0078	.0352	.0860	.1468	.1963	.2178	.2066	.1701	.1222
	6	.0000	.0013	.0093	.0322	.0734	.1262	.1759	.2066	.2088	.1833
	7	.0000	.0002	.0019	.0092	.0280	.0618	.1082	.1574	.1952	.2095
	8	.0000	.0000	.0003	.0020	.0082	.0232	.0510	.0918	.1398	.1833
	9	.0000	.0000	.0000	.0003	.0018	.0066	.0183	.0408	.0762	.1222
	10	.0000	.0000	.0000	.0000	.0003	.0014	.0049	.0136	.0312	.0611
	11	.0000	.0000	.0000	.0000	.0000	.0002	.0010	.0033	.0093	.0222
	12	.0000	.0000	.0000	.0000	.0000	.0000	.0001	.0005	.0019	.0056
	13	.0000	.0000	.0000	.0000	.0000	.0000	.0000	.0001	.0002	.0009
	14	.0000	.0000	.0000	.0000	.0000	.0000	.0000	.0000	.0000	.0001
15	0	.4633	.2059	.0874	.0352	.0134	.0047	.0016	.0005	.0001	.0000
	1	.3658	.3432	.2312	.1319	.0668	.0305	.0126	.0047	.0016	.0005
	2	.1348	.2669	.2856	.2309	.1559	.0916	.0476	.0219	.0090	.0032
	3	.0307	.1285	.2184	.2501	.2252	.1700	.1110	.0634	.0318	.0139
	4	.0049	.0428	.1156	.1876	.2252	.2186	.1792	.1268	.0780	.0417
	5	.0006	.0105	.0449	.1032	.1651	.2061	.2123	.1859	.1404	.0916
	6	.0000	.0019	.0132	.0430	.0917	.1472	.1906	.2066	.1914	.1527
	7	.0000	.0003	.0030	.0138	.0393	.0811	.1319	.1771	.2013	.1964
	8	.0000	.0000	.0005	.0035	.0131	.0348	.0710	.1181	.1647	.1964
	9	.0000	.0000	.0001	.0007	.0034	.0116	.0298	.0612	.1048	.1527

Table 1 (continued)

n	k	.05	.10	.15	.20	.25	.30	.35	.40	.45	.50
						p					
15	10	.0000	.0000	.0000	.0001	.0007	.0030	.0096	.0245	.0515	.0916
	11	.0000	.0000	.0000	.0000	.0001	.0006	.0024	.0074	.0191	.0417
	12	.0000	.0000	.0000	.0000	.0000	.0001	.0004	.0016	.0052	.0139
	13	.0000	.0000	.0000	.0000	.0000	.0000	.0001	.0003	.0010	.0032
	14	.0000	.0000	.0000	.0000	.0000	.0000	.0000	.0000	.0001	.0005
	15	.0000	.0000	.0000	.0000	.0000	.0000	.0000	.0000	.0000	.0000
16	0	.4401	.1853	.0743	.0281	.0100	.0033	.0010	.0003	.0001	.0000
	1	.3706	.3294	.2097	.1126	.0535	.0228	.0087	.0030	.0009	.0002
	2	.1463	.2745	.2775	.2111	.1336	.0732	.0353	.0150	.0056	.0018
	3	.0359	.1423	.2235	.2463	.2079	.1465	.0888	.0468	.0215	.0085
	4	.0061	.0514	.1311	.2001	.2252	.2040	.1553	.1014	.0572	.0278
	5	.0008	.0137	.0555	.1201	.1802	.2099	.2008	.1623	.1123	.0667
	6	.0001	.0028	.0180	.0550	.1101	.1649	.1982	.1983	.1684	.1222
	7	.0000	.0004	.0045	.0197	.0524	.1010	.1524	.1889	.1969	.1746
	8	.0000	.0001	.0009	.0055	.0197	.0487	.0923	.1417	.1812	.1964
	9	.0000	.0000	.0001	.0012	.0058	.0185	.0442	.0840	.1318	.1746
	10	.0000	.0000	.0000	.0002	.0014	.0056	.0167	.0392	.0755	.1222
	11	.0000	.0000	.0000	.0000	.0002	.0013	.0049	.0142	.0337	.0667
	12	.0000	.0000	.0000	.0000	.0000	.0002	.0011	.0040	.0115	.0278
	13	.0000	.0000	.0000	.0000	.0000	.0000	.0002	.0008	.0029	.0085
	14	.0000	.0000	.0000	.0000	.0000	.0000	.0000	.0001	.0005	.0018
	15	.0000	.0000	.0000	.0000	.0000	.0000	.0000	.0000	.0001	.0002
	16	.0000	.0000	.0000	.0000	.0000	.0000	.0000	.0000	.0000	.0000
17	0	.4181	.1668	.0631	.0225	.0075	.0023	.0007	.0002	.0000	.0000
	1	.3741	.3150	.1893	.0957	.0426	.0169	.0060	.0019	.0005	.0001
	2	.1575	.2800	.2673	.1914	.1136	.0581	.0260	.0102	.0035	.0010
	3	.0415	.1556	.2359	.2393	.1893	.1245	.0701	.0341	.0144	.0052
	4	.0076	.0605	.1457	.2093	.2209	.1868	.1320	.0796	.0411	.0182
	5	.0010	.0175	.0668	.1361	.1914	.2081	.1849	.1379	.0875	.0472
	6	.0001	.0039	.0236	.0680	.1276	.1784	.1991	.1839	.1432	.0944
	7	.0000	.0007	.0065	.0267	.0668	.1201	.1685	.1927	.1841	.1484
	8	.0000	.0001	.0014	.0084	.0279	.0644	.1134	.1606	.1883	.1855
	9	.0000	.0000	.0003	.0021	.0093	.0276	.0611	.1070	.1540	.1855

Table 1 (continued)

n	k	.05	.10	.15	.20	.25	.30	.35	.40	.45	.50
						p					
17	10	.0000	.0000	.0000	.0004	.0025	.0095	.0263	.0571	.1008	.1484
	11	.0000	.0000	.0000	.0001	.0005	.0026	.0090	.0242	.0525	.0944
	12	.0000	.0000	.0000	.0000	.0001	.0006	.0024	.0081	.0215	.0472
	13	.0000	.0000	.0000	.0000	.0000	.0001	.0005	.0021	.0068	.0182
	14	.0000	.0000	.0000	.0000	.0000	.0000	.0001	.0004	.0016	.0052
	15	.0000	.0000	.0000	.0000	.0000	.0000	.0000	.0001	.0003	.0010
	16	.0000	.0000	.0000	.0000	.0000	.0000	.0000	.0000	.0000	.0001
	17	.0000	.0000	.0000	.0000	.0000	.0000	.0000	.0000	.0000	.0000
18	0	.3972	.1501	.0536	.0180	.0056	.0016	.0004	.0001	.0000	.0000
	1	.3763	.3002	.1704	.0811	.0338	.0126	.0042	.0012	.0003	.0001
	2	.1683	.2835	.2556	.1723	.0958	.0458	.0190	.0069	.0022	.0006
	3	.0473	.1680	.2406	.2297	.1704	.1046	.0547	.0246	.0095	.0031
	4	.0093	.0700	.1592	.2153	.2130	.1681	.1104	.0614	.0291	.0117
	5	.0014	.0218	.0787	.1507	.1988	.2017	.1664	.1146	.0666	.0327
	6	.0002	.0052	.0301	.0816	.1436	.1873	.1941	.1655	.1181	.0708
	7	.0000	.0010	.0091	.0350	.0820	.1376	.1792	.1892	.1657	.1214
	8	.0000	.0002	.0022	.0120	.0376	.0811	.1327	.1734	.1864	.1669
	9	.0000	.0000	.0004	.0033	.0139	.0386	.0794	.1284	.1694	.1855
	10	.0000	.0000	.0001	.0008	.0042	.0149	.0385	.0771	.1248	.1669
	11	.0000	.0000	.0000	.0001	.0010	.0046	.0151	.0374	.0742	.1214
	12	.0000	.0000	.0000	.0000	.0002	.0012	.0047	.0145	.0354	.0708
	13	.0000	.0000	.0000	.0000	.0000	.0002	.0012	.0045	.0134	.0327
	14	.0000	.0000	.0000	.0000	.0000	.0000	.0002	.0011	.0039	.0117
	15	.0000	.0000	.0000	.0000	.0000	.0000	.0000	.0002	.0009	.0031
	16	.0000	.0000	.0000	.0000	.0000	.0000	.0000	.0000	.0001	.0006
	17	.0000	.0000	.0000	.0000	.0000	.0000	.0000	.0000	.0000	.0001
	18	.0000	.0000	.0000	.0000	.0000	.0000	.0000	.0000	.0000	.0000
19	0	.3774	.1351	.0456	.0144	.0042	.0011	.0003	.0001	.0000	.0000
	1	.3774	.2852	.1529	.0685	.0268	.0093	.0029	.0008	.0002	.0000
	2	.1787	.2852	.2428	.1540	.0803	.0358	.0138	.0046	.0013	.0003
	3	.0533	.1796	.2428	.2182	.1517	.0869	.0422	.0175	.0062	.0018
	4	.0112	.0798	.1714	.2182	.2023	.1491	.0909	.0467	.0203	.0074

Table 1 (continued)

n	*k*	.05	.10	.15	.20	.25	.30	.35	.40	.45	.50
19	5	.0018	.0266	.0907	.1636	.2023	.1916	.1468	.0933	.0497	.0222
	6	.0002	.0069	.0374	.0955	.1574	.1916	.1844	.1451	.0949	.0518
	7	.0000	.0014	.0122	.0443	.0974	.1525	.1844	.1797	.1443	.0961
	8	.0000	.0002	.0032	.0166	.0487	.0981	.1489	.1797	.1771	.1442
	9	.0000	.0000	.0007	.0051	.0198	.0514	.0980	.1464	.1771	.1762
	10	.0000	.0000	.0001	.0013	.0066	.0220	.0528	.0976	.1449	.1762
	11	.0000	.0000	.0000	.0003	.0018	.0077	.0233	.0532	.0970	.1442
	12	.0000	.0000	.0000	.0000	.0004	.0022	.0083	.0237	.0529	.0961
	13	.0000	.0000	.0000	.0000	.0001	.0005	.0024	.0085	.0233	.0518
	14	.0000	.0000	.0000	.0000	.0000	.0001	.0006	.0024	.0082	.0222
	15	.0000	.0000	.0000	.0000	.0000	.0000	.0001	.0005	.0022	.0074
	16	.0000	.0000	.0000	.0000	.0000	.0000	.0000	.0001	.0005	.0018
	17	.0000	.0000	.0000	.0000	.0000	.0000	.0000	.0000	.0001	.0003
	18	.0000	.0000	.0000	.0000	.0000	.0000	.0000	.0000	.0000	.0000
	19	.0000	.0000	.0000	.0000	.0000	.0000	.0000	.0000	.0000	.0000
20	0	.3585	.1216	.0388	.0115	.0032	.0008	.0002	.0000	.0000	.0000
	1	.3774	.2702	.1368	.0576	.0211	.0068	.0020	.0005	.0001	.0000
	2	.1887	.2852	.2293	.1369	.0669	.0278	.0100	.0031	.0008	.0002
	3	.0596	.1901	.2428	.2054	.1339	.0716	.0323	.0123	.0040	.0011
	4	.0133	.0898	.1821	.2182	.1897	.1304	.0738	.0350	.0139	.0046
	5	.0022	.0319	.1028	.1746	.2023	.1789	.1272	.0746	.0365	.0148
	6	.0003	.0089	.0454	.1091	.1686	.1916	.1712	.1244	.0746	.0370
	7	.0000	.0020	.0160	.0545	.1124	.1643	.1844	.1659	.1221	.0739
	8	.0000	.0004	.0046	.0222	.0609	.1144	.1614	.1797	.1623	.1201
	9	.0000	.0001	.0011	.0074	.0271	.0654	.1158	.1597	.1771	.1602
	10	.0000	.0000	.0002	.0020	.0099	.0308	.0686	.1171	.1593	.1762
	11	.0000	.0000	.0000	.0005	.0030	.0120	.0336	.0710	.1185	.1602
	12	.0000	.0000	.0000	.0001	.0008	.0039	.0136	.0355	.0727	.1201
	13	.0000	.0000	.0000	.0000	.0002	.0010	.0045	.0146	.0366	.0739
	14	.0000	.0000	.0000	.0000	.0000	.0002	.0012	.0049	.0150	.0370
	15	.0000	.0000	.0000	.0000	.0000	.0000	.0003	.0013	.0049	.0148
	16	.0000	.0000	.0000	.0000	.0000	.0000	.0000	.0003	.0013	.0046
	17	.0000	.0000	.0000	.0000	.0000	.0000	.0000	.0000	.0002	.0011
	18	.0000	.0000	.0000	.0000	.0000	.0000	.0000	.0000	.0000	.0002
	19	.0000	.0000	.0000	.0000	.0000	.0000	.0000	.0000	.0000	.0000
	20	.0000	.0000	.0000	.0000	.0000	.0000	.0000	.0000	.0000	.0000

Table 2

Areas under the Normal Distribution

b*	.00	.01	.02	.03	.04	.05	.06	.07	.08	.09
0.0	.0000	.0040	.0080	.0120	.0160	.0199	.0239	.0279	.0319	.0359
0.1	.0398	.0438	.0478	.0517	.0557	.0596	.0636	.0675	.0714	.0753
0.2	.0793	.0832	.0871	.0910	.0948	.0987	.1026	.1064	.1103	.1141
0.3	.1179	.1217	.1255	.1293	.1331	.1368	.1406	.1443	.1480	.1517
0.4	.1554	.1591	.1628	.1664	.1700	.1736	.1772	.1808	.1844	.1879
0.5	.1915	.1950	.1985	.2019	.2054	.2088	.2123	.2157	.2190	.2224
0.6	.2257	.2291	.2324	.2357	.2389	.2422	.2454	.2486	.2517	.2549
0.7	.2580	.2611	.2642	.2673	.2704	.2734	.2764	.2794	.2823	.2852
0.8	.2881	.2910	.2939	.2967	.2995	.3023	.3051	.3078	.3106	.3133
0.9	.3159	.3186	.3212	.3238	.3264	.3289	.3315	.3340	.3365	.3389
1.0	.3413	.3438	.3461	.3485	.3508	.3531	.3554	.3577	.3599	.3621
1.1	.3643	.3665	.3686	.3708	.3729	.3749	.3770	.3790	.3810	.3830
1.2	.3849	.3869	.3888	.3907	.3925	.3944	.3962	.3980	.3997	.4015
1.3	.4032	.4049	.4066	.4082	.4099	.4115	.4131	.4147	.4162	.4177
1.4	.4192	.4207	.4222	.4236	.4251	.4265	.4279	.4292	.4306	.4319
1.5	.4332	.4345	.4357	.4370	.4382	.4394	.4406	.4418	.4429	.4441
1.6	.4452	.4463	.4474	.4484	.4495	.4505	.4515	.4525	.4535	.4545
1.7	.4554	.4564	.4573	.4582	.4591	.4599	.4608	.4616	.4625	.4633
1.8	.4641	.4649	.4656	.4664	.4671	.4678	.4686	.4693	.4699	.4706
1.9	.4713	.4719	.4726	.4732	.4738	.4744	.4750	.4756	.4761	.4767
2.0	.4772	.4778	.4783	.4788	.4793	.4798	.4803	.4808	.4812	.4817
2.1	.4821	.4826	.4830	.4834	.4838	.4842	.4846	.4850	.4854	.4857
2.2	.4861	.4864	.4868	.4871	.4875	.4878	.4881	.4884	.4887	.4890
2.3	.4893	.4896	.4898	.4901	.4904	.4906	.4909	.4911	.4913	.4916
2.4	.4918	.4920	.4922	.4925	.4927	.4929	.4931	.4932	.4934	.4936
2.5	.4938	.4940	.4941	.4943	.4945	.4946	.4948	.4949	.4951	.4952
2.6	.4953	.4955	.4956	.4957	.4959	.4960	.4961	.4962	.4963	.4964
2.7	.4965	.4966	.4967	.4968	.4969	.4970	.4971	.4972	.4973	.4974
2.8	.4974	.4975	.4976	.4977	.4977	.4978	.4979	.4979	.4980	.4981
2.9	.4981	.4982	.4982	.4983	.4984	.4984	.4985	.4985	.4986	.4986
3.0	.4987	.4987	.4987	.4988	.4988	.4989	.4989	.4989	.4990	.4990

* For $b > 3$, the area under the curve is essentially 0.5.

Appendix 4

ANSWERS TO EXERCISES

Chapter 1

1. (a) true; (b) false; (c) true; (d) false; (e) true; (f) true; (g) false; (h) true
2. $\{26, 27, 28, \ldots, 50\}$, $\{1, 2, 3, \ldots, 9, 36, 37, \ldots, 50\}$, $\{1, 3, \ldots, 49\}$
3. $\{10, 11, 12, \ldots, 25\}$, $\{2, 4, 6, \ldots, 24\}$, $\{26, 28, 30, \ldots, 50\}$
4. $\{1, 2, 3, \ldots, 35\}$, $\{1, 2, 3, \ldots, 26, 28, 30, \ldots, 50\}$,
   $\{2, 4, 6, \ldots, 26, 27, 28, \ldots, 50\}$, $\{1, 2, \ldots, 9, 11, 13, \ldots, 35, 36, \ldots, 50\}$
   $\{2, 4, 6, 8, 10, 11, 12, 13, \ldots, 35\}$
5. $\{2, 4, 6, 8, 10, 11, 12, \ldots, 25\}$
6. $\{1, 2, 3, \ldots, 25, 26, 28, 30, 32, 34\}$
7. $\{36, 37, \ldots, 50\}$, $\{1, 2, 3, \ldots, 9, 26, 27, 28, \ldots, 50\}$

9. $\{4, 1, 0, 9, 16\}$
10. $\{\pm 1, \pm\sqrt{2}, \pm\sqrt{3}, \pm 2, \pm\sqrt{5}, \pm\sqrt{6}\}$
11. $\{8, 6, 4, 2, 0, -2, -4\}$
12. $\{(1, 7), (3, 6), (5, 5), (7, 4), (9, 3), (11, 2), (13, 1)\}$
13. $\varnothing$, {#}; $\varnothing$, {@}, {\$}, {@, \$}; $\varnothing$, {%}, {¢}, {&}, {%, ¢}, {%, &},
    {¢, &}, {%, ¢, &}; 2^n
14. 30; the set of students who passed both exams is $A \cap B$.
    "The set of students who passed at least one exam."
15. 19

Chapter 2

Section 2.1

1. (a) $y = 2x + 4$; (b) $y = -3x$; (c) $y = -2$; (d) $y = (1/2)x - 1/2$;
   (e) $y = (-1/4)x + 3$
2. (a) $3x + y = 11$; (b) $x - 2y + 2 = 0$; (c) $6y = x$; (d) $x + 4y + 6 = 0$
3. 2/5, 3, $-6/5$
4. $2x - 5y + 37 = 0$
5. 121/24
6. $x = -11/2$
7. $y = -7$, $y = 1$
8. (a) $14x + 2y = 3$; (b) $24x + y + 9 = 0$; (c) $2x + y = 0$; (d) $y = -5$;
   (e) $x = 1/2$; (f) $4x - 7y + 28 = 0$

10. $3x - 5y + 15 = 0$
11. $2x + 5y = 20$, $2x - 5y + 20 = 0$
12. $y = -5x + 20$, $y = -5x - 20$
13. $\sqrt{193}$
14. No

Section 2.2

4. (a) $3x + y \geq 11$; (b) $5x + 3y \geq 4$; (c) $y \leq 7$ and $x \geq -3$;
   (d) $-1/4 \leq x \leq 3$ and $y \leq 6$

Section 2.3

2. $x - 2y \geq -11$, $7x + 2y \leq 35$, $4x + 9y \geq 20$, $8x - y \geq 2$
3. $x \geq 0$, $y \geq 0$, $x + 5y \leq 50$, $3x + y \leq 24$, $6x + 7y \leq 42$
4. (a) max $P = 95$, min $P = 13$, max $C = -4$, min $C = -102$; (b) max $P = 70$, min $P = -34$, max $C = 24$, min $C = -67$; (c) no maximum for P, min $P = 108/11$, no maximum or minimum for C; (d) max $P = 36$, min $P = 11$, max $C = -3$, min $C = -38$; (e) max $P = 86.5$, min $P = 0$, max $C = 46$, min $C = -70$
5. 8.7 boxes of the first kind, 5.2 boxes of the second kind. Profit is $5.04.
6. 10 of the first type, 14 of the second type. Profit is $62.
7. 12 of the first type, 6 of the second type. Profit is $28. Price should be between $.60 and $1.50.

Section 2.4

1. 36
2. 40 suits, no dresses, income is $3200.
3. 5 commercial and 25 residential pickups at a total cost of $875
4. 20/3 grams of A and 20/3 grams of B produce 1000 calories. Since the conditions say "more of B than of A," strictly speaking, there is no perfect solution. Roughly 6.6 grams of A, 6.7 grams of B is the best possible.
5. No change
8. $M = (11/3, 4/3)$; $2x - 7y + 2 = 0$; $8x + 5y - 36 = 0$; $5x - y - 17 = 0$
9. $f = 4/5$, $g = 1/5$

Chapter 3

Section 3.1

1. $(9, -1)$
2. $(-1, 3, 2)$
3. $(-2, 7, 4)$
4. $(-1, 1/2, -1/3)$
5. $(-2, 1, 3, 5/2)$
6. $x = 8 - k$, $y = 9 - k$, $z = 6 - k$, $w = k$, for any value of k
7. $(13/37, 19/37, 5/37)$
8. $(19/45, 1/9, 7/15)$

Section 3.2

1. $(2/3, 1/3)$
2. $f = (1 - 7n)/5$, $g = (12 + 61n)/60$, $h = (36 - 37n)/60$, $k = n$, for any value of n
3. Inconsistent
4. Inconsistent
5. $x = (13 + n)/4$, $y = (1 + 5n)/4$, $z = n$, for any n
6. Inconsistent
7. $x = 21/10$, $y = 13/10$

8. Inconsistent
9. $x = (17 + n)/3$, $y = (4n - 1)/3$, $z = n$, $w = -2$, for any n
10. $x = (18n - 7m - 6)/3$, $y = (21 - m)/3$, $z = m$, $w = n$, where m and n are any numbers

Section 3.3

19. $x = (23 - 3n)/3$, $y = 5/3$, $z = n$, for any n

Chapter 4

Section 4.1

1. 25/6; 0
2. {double 1, 3, double 2, 4 (no double), 5, double 3, 6 (no double), 7, double 4, 8 (no double), 9, double 5, 10 (no double), 11, double 6}
3. {1, 2, 3, 4, 5, 6, 7, 8, 9, 10}
4. (a) {(1, 2), (3, 4), (5, 6), (7, 8), (9, 10)}, many other possibilities; (b) {(1, 2, 3), 4, (5, 6), 7, (8, 9, 10)}, many other possibilities
5. $V = \{a, e, i, o, u\}$, $E = \{e, n, t, v\}$, $\{e\} = V \cap E$; $E \cup V$; $E \cap V'$
6. $\varnothing$, $\{R\}$, $\{G\}$, $\{B\}$, $\{Y\}$, $\{R, G\}$, $\{R, B\}$, $\{R, Y\}$, $\{G, B\}$, $\{G, Y\}$, $\{B, Y\}$, $\{R, G, B\}$, $\{R, G, Y\}$, $\{R, B, Y\}$, $\{G, B, Y\}$, S
7. {(1, 1), (1, 2), (1, 3), (1, 4), (1, 5), (1, 6), (2, 1), (2, 2), (2, 3), (2, 4), (2, 5), (2, 6), (3, 1), (3, 2), (3, 3), (3, 4), (3, 5), (3, 6), (4, 1), (4, 2), (4, 3), (4, 4), (4, 5), (4, 6), (5, 1), (5, 2), (5, 3), (5, 4), (5, 5), (5, 6), (6, 1), (6, 2), (6, 3), (6, 4), (6, 5), (6, 6)}
8. {first try, second try, third try, fourth try, last try, never}
9. {first try, second try}′, {first try, second try, third try}
   {first try, never}, {second try, third try, fourth try}
10. {(R, R), (R, S), (R, C), (S, R), (S, S), (S, C), (C, R), (C, S), (C, C)}
    {(R, R), (R, S), (R, C), (S, R), (C, R)}, {(R, S), (R, C), (S, R), (C, R)}
    {(R, R), (S, S), (C, C)}, {(R, R), (S, S), (C, C)}′,
    {(S, R), (S, S), (S, C)}, {(S, R), (S, C), (R, S), (C, S)}

Section 4.2

2. 1/2, 3/5, 3/10, 4/5
3. 5/26, 2/13, 4/13, 3/26
4. In the order listed in the answer to Exercise 6, Section 4.1: 0, 1/2, 1/4, 1/8, 1/8, 3/4, 5/8, 5/8, 3/8, 3/8, 1/4, 7/8, 7/8, 3/4, 1/2, 1
5. 4/9, 7/9, 7/18, 5/9
6. 71%
7. 0.3, 0.4, 0.4, 0.1
8. The conditions are incompatible.
9. 1/36, 0, 11/36
10. 0.6

Section 4.3

1. See Section 4.1, Exercise 7. The probabilities are 1/36, 2/36, 3/36, 4/36, 5/36, 6/36, 5/36, 4/36, 3/36, 2/36, 1/36.

2. HHH, HHT, HTH, THH, HTT, THT, TTH, TTT (a) 1/8; (b) 1/2; (c) 1/2; (d) 1/2; (e) 3/4; (f) 1/4; (g) 7/8
5. 9/47
6. 4/47
7. 4/47, 8/47, 15/47

Section 4.4

1. $p(x) = 1/2$, for $x = 0, 1$
2. $p(0) = 1/16$, $p(1) = 1/4$, $p(2) = 3/8$, $p(3) = 1/4$, $p(4) = 1/16$
3. 5/16, 7/8
4. $p(0) = 0.15$, $p(1) = 0.49$, $p(2) = 0.24$, $p(3) = 0.12$
5. $p(0) = 0.12$, $p(1) = 0.24$, $p(2) = 0.49$, $p(3) = 0.15$
6. $p(0) = 3/8$, $p(1) = 1/2$, $p(4) = 1/8$
7. $p_X(1) = 1/15$, $p_X(2) = 1/3$, $p_X(3) = 3/5$, $p_Y(0) = 1/5$, $p_Y(1) = 1/5$, $p_Y(2) = 4/15$, $p_Y(3) = 1/3$
8.

z	−5	−4	−3	−2	−1	0	1	2	3	4	5
$p(z)$	1/15	1/15	1/15	1/15	2/15	3/15	2/15	1/15	1/15	1/15	1/15

9. $p(0) = 3/15$, $p(1) = 8/15$, $p(2) = 4/15$
10. For $(X - 1)^2$: $p(0) = 1/15$, $p(1) = 1/3$, $p(4) = 3/5$
    For $(Y - 1)^2$: $p(0) = 1/5$, $p(1) = 7/15$, $p(4) = 1/3$
    For $(Z - 1)^2$: $p(0) = 2/15$, $p(1) = 4/15$, $p(4) = 3/15$, $p(9) = 2/15$, $p(16) = 2/15$, $p(25) = 1/15$, $p(36) = 1/15$
    For $(W - 1)^2$: $p(0) = 8/15$, $p(1) = 7/15$

Section 4.5

1. 1/2, 1, 3/2, 2, . . . ; $n/2$ for n flips
2. 1.33
3. 1.67
4. 1
5. $E(X) = 38/15$, $E(Y) = 26/15$
6. 0
7. 16/15
8. $E[(X - 1)^2] = 41/15$, $E[(Y - 1)^2] = 9/5$, $E[(Z - 1)^2] = 127/15$, $E[(W - 1)^2] = 7/15$
9. 20¢
10. −1/19, i.e., you average a little more than a 5¢ loss on each $1.00 bet.
11. (a) 13; (b) −2
12. (a) 4; (b) −3, 1; (c) impossible; (d) any number
13. (a) (8/3, −1/3); (b) (30, −8)
14. The square of the distance from the origin to (x, y)

Section 4.6

1. 35/6
4. −5
5. 16

6. 2, 4/3
7. −13
8. (−4, 5, 1)
9. 18
10. 0, 5
11. 8, −27, 17
13. $(2\sqrt{5}/5, \sqrt{5}/5), (-2\sqrt{5}/5, -\sqrt{5}/5)$
14. $(k, k, 0)$, where k is any number

Section 4.7

1. 35/12
2. 0.7611
3. 0.7611
4. 3/2
5. Var $(X) = 86/225$, Var $(Y) = 284/225$
6. 112/15
7. 104/225
8. Var $[(X-1)^2] = 554/225$, Var $[(Y-1)^2] = 64/25$, Var $[(Z-1)^2] = 23576/225$, Var $[(W-1)^2] = 56/225$
9. 17/3, 35/9
11. 95/13, 1440/169

Section 4.8

2. 9
3. (−7, 13, −1, 9)
4. 1, 10, 9
5. 0, 3, −4
6. 11, −36, 41
7. (a) (4/7, 3/7), $\begin{pmatrix} 5/7 \\ 2/7 \end{pmatrix}$, $v = -1/7$

   (b) (0, 1), $\begin{pmatrix} 1 \\ 0 \end{pmatrix}$, $v = 4$. See Section 4.9.

   (c) (3/4, 1/4), $\begin{pmatrix} 1/2 \\ 1/2 \end{pmatrix}$, $v = -3/2$
8. $\begin{pmatrix} 3 & -1 \\ -18 & 6 \end{pmatrix}$, (6/7, 1/7), $\begin{pmatrix} 1/4 \\ 3/4 \end{pmatrix}$, $v = 0$
9. Joe should pass two out of seven times, for an average gain of 50/7 yards. His opponents should prepare for a pass on nine out of every fourteen plays.

Section 4.9

2. (a) (1, 0), $\begin{pmatrix} 1 \\ 0 \end{pmatrix}$, $v = -1$

   (b) (0, 1), $\begin{pmatrix} 0 \\ 1 \\ 0 \end{pmatrix}$, $v = 1$

(c) $(0, 1, 0), \begin{pmatrix} 0 \\ 1 \\ 0 \\ 0 \end{pmatrix}, v = -1$

3. (a) Player I plays either Row 3 or Row 4, Player II plays Column 4, $v = 5$.

   (b) $(1/2, 1/2, 0), \begin{pmatrix} 1/2 \\ 1/2 \\ 0 \end{pmatrix}, v = 3/2$

4. (a) $(0, 1/2, 1/2), \begin{pmatrix} 1/6 \\ 0 \\ 5/6 \end{pmatrix}, v = 5/2$

   (b) Player I plays Row 2, Player II plays Column 2 or 3, $v = 4$.

5. Player I plays the black 3, Player II plays the black 1, $v = 3$.
6. This is a fair game. Both players should play 1 or 2 with equal probability and never play 3. The value of the game is zero.

Section 4.10

1. (a) $(1/2, 1/2), \begin{pmatrix} 0 \\ 1/3 \\ 0 \\ 2/3 \end{pmatrix}, v = 1$

   (b) $(5/18, 1/2, 0, 2/9), \begin{pmatrix} 10/27 \\ 11/27 \\ 2/9 \end{pmatrix}, v = 4/9$

3. Player I plays black 3 and red 3 each with probability 1/2. Player II plays red 1 and black 2 with probabilities 2/3 and 1/3 respectively, or red 1 and black 3 with probabilities 3/4 and 1/4 respectively (or any weighted average of these two strategies). The value of the game is 3.
4. Joe plays 1, 2 and 3 with probabilities 2/3, 1/6 and 1/6 respectively. Sam does the same. The value of the game is $-1/3$.
5. (a) $(7/9, 0, 0, 0, 2/9), \begin{pmatrix} 4/9 \\ 5/9 \end{pmatrix}, v = 53/9$

   (b) $(0, 1, 0, 0), \begin{pmatrix} 0 \\ 1 \end{pmatrix}, v = 5$

   (c) $(2/3, 1/3), \begin{pmatrix} 0 \\ 1/2 \\ 1/2 \\ 0 \end{pmatrix}, v = 2$

   (d) $(1, 0, 0), \begin{pmatrix} 0 \\ 1 \\ 0 \end{pmatrix}, v = 5$

   (e) $(1/5, 0, 4/5), \begin{pmatrix} 4/5 \\ 1/5 \\ 0 \end{pmatrix}, v = 37/5$

(f) $(0, 1/5, 4/5), \begin{pmatrix} 0 \\ 1/5 \\ 4/5 \\ 0 \end{pmatrix}, v = 16/5$

(g) $(0, 0, 1/2, 1/2), \begin{pmatrix} 0 \\ 0 \\ 1/8 \\ 7/8 \end{pmatrix}, v = 11/2$

Chapter 5

Section 5.1

1. (a) 1/3; (b) 2/7; (c) 2/7
2. (a) 0.6976; (b) 0.648; (c) 0.2016
3. 11/850, 33/66,640, 33/16,660
4. 8/19, 3/8
5. 33/80, 49/99, 4/49
6. 4/47
7. (a) 2/5; (b) 1/3; (c) 2/15
8. 1/3
9. (a) 1/8; (b) 5/24; (c) 11/24; (d) 3/8
10. 1/36
11. 5/6, 5/7

Section 5.2

1. $P(U_1 \mid W) = 3/4$, $P(U_2 \mid W) = 1/4$
2. 561/1072, 561/1072
3. 171/179
4. 2/3
5. $P(\text{small} \mid \text{rotten}) = 1/8$, $P(\text{medium} \mid \text{rotten}) = 3/8$, $P(\text{large} \mid \text{rotten}) = 1/2$
6. 4/13, 8/13, 1/13
7. 8/17, 8/17, 1/17
8. (a) 2/9, 0, 7/9; (b) 4/53, 0, 49/53

Section 5.3

1. 0.56
2. 11/15
3. 0.5
5. 0.7, 0.2
6. No
7. Without replacement, 1/84; with replacement, 1/27
8. 8/27
9. 65/81
10. 3/4, 1/3, 1/4, 4/15, 29/30
11. 2/3; no

Section 5.4

1. (b) 1; (c) 1/2; (d) 4, 0
2. (b) 0; (c) 1/2; (d) 4.5, 2/3
3. (b) 1/2; (c) 1/2; (d) 4.25, 1/3
4. 1/8, 19/72, 13/36, 1/4; 125/72, 26/47
5. For the first model: 1/4, 1/4, 1/4, 1/4; 1.5; 2/3. For the second model: 1/8, 7/32, 21/64, 21/64; 119/64; 27/47
6. 3/20, 2/5
7. 147/512, 315/512
8. 13/48, 25/48
9. (a) 19/48; (b) 57/157; (c) 43/48

Section 5.5

1. (a) Not a transition matrix. $\mathbf{P}^2 = \begin{pmatrix} 1/6 & 1/3 \\ 2/9 & 11/18 \end{pmatrix}$

   (b) Transition matrix. $\mathbf{P}^2 = \begin{pmatrix} 1 & 0 \\ 0 & 1 \end{pmatrix}$

   (c) Not a transition matrix. $\mathbf{P}^2 = \begin{pmatrix} .22 & .49 & .29 \\ -.13 & .76 & .37 \\ -.05 & .65 & .40 \end{pmatrix}$

   (d) Transition matrix. $\mathbf{P}^2 = \begin{pmatrix} .25 & .11 & .64 \\ 0 & 1 & 0 \\ .24 & .03 & .73 \end{pmatrix}$
2. Hamburger probability is 44%; chicken salad, 56%.
3. 0.195 (0.4280, 0.2805, 0.2915)
4. (b) $P(\text{Jo}) = 13/36$, $P(\text{Ed}) = 14/36$, $P(\text{Howie}) = 9/36$; (c) $P(\text{Jo}) = 305/864$, $P(\text{Ed}) = 322/864$, $P(\text{Howie}) = 237/864$
5. (5/13, 8/13)
6. $7/27 \approx 25.9\%$
7. 53/80
8. (0.2625, 0.2125, 0.3125, 0.2125)
9. (b) 0.44, 0.576
10. (b) 0.4; (c) 0.7
11. (a) $\mathbf{P}^3 = \begin{pmatrix} 1/27 & 6/27 & 20/27 \\ 0 & 1/27 & 26/27 \\ 0 & 0 & 1 \end{pmatrix}$

    (b) $\mathbf{P}^3 = \begin{pmatrix} 11/32 & 21/32 \\ 21/64 & 43/64 \end{pmatrix}$

    (c) $\mathbf{P}^3 = \begin{pmatrix} .2485 & .2205 & .3105 & .2205 \\ .2925 & .1900 & .3275 & .1900 \\ .2275 & .21375 & .3450 & .21375 \\ .2275 & .21375 & .3450 & .21375 \end{pmatrix}$

    (d) $\mathbf{P}^3 = \begin{pmatrix} .688 & .312 \\ .624 & .376 \end{pmatrix}$ (e) $\mathbf{P}^3 = \begin{pmatrix} .544 & .456 \\ .608 & .392 \end{pmatrix}$

Section 5.6

1. (a) Not a transition matrix
   (b) $\mathbf{w} = (1/2, 1/2)$, not regular
   (c) Not a transition matrix
   (d) $\mathbf{w} = (0, 1, 0)$, not regular
2. $50/201 = 24.88\%$
3. $(4/11, 4/11, 3/11)$
4. $\mathbf{w} = (0, 0, 1)$, for example,
$$\mathbf{P}^8 = \begin{pmatrix} .0002 & .0024 & .9974 \\ 0 & .0002 & .9998 \\ 0 & 0 & 1 \end{pmatrix}$$
5. 1/3 laborers, 2/3 professionals
6. $(5/19, 4/19, 6/19, 4/19)$
7. 2/3 smoke brand C, 1/3 do not
8. 4/7 easy, 3/7 difficult
9. Regular, $(1/3, 1/4, 1/4, 1/6)$

Section 5.7

1. (b) $\begin{pmatrix} -2 & 7 \\ -15 & 25 \\ -3 & 5 \\ -20 & 37 \end{pmatrix}$ (c) $\begin{pmatrix} -2 & 9 \\ -9 & 15 \\ 0 & -5 \\ -23 & 49 \end{pmatrix}$
   (d) $\begin{pmatrix} 1 & -1 \\ -6 & 15 \\ -3 & 11 \\ 4 & -5 \end{pmatrix}$ (e) $\begin{pmatrix} 1 & -31 \\ 20 & -15 \\ 9 & 40 \\ 55 & -176 \end{pmatrix}$
2. (c) $\mathbf{D}^2 + \mathbf{DE} + \mathbf{ED} + \mathbf{E}^2$
4. $\begin{pmatrix} 3m & 3n \\ 2m & 2n \end{pmatrix}$ for any numbers m and n
5. $x = 3, 9$
6. Not possible
7. (a) $\begin{pmatrix} -12 & 6 \\ 10 & -28 \\ 35 & -29 \\ -47 & 35 \end{pmatrix}$ (b) $\begin{pmatrix} -6 & 20 \\ -25 & 70 \\ -5 & 14 \\ -36 & 104 \end{pmatrix}$
8. $\mathbf{GH} = 25$, $\mathbf{HG} = \begin{pmatrix} -1 & 2 & 4 \\ -3 & 6 & 12 \\ -5 & 10 & 20 \end{pmatrix}$
10. $\begin{pmatrix} 7 & 4 \\ 37 & 29 \\ 32 & 25 \\ 5 & 15 \end{pmatrix}$

Section 5.8

1. $\begin{pmatrix} 5 & -11 \\ 5 & -9 \end{pmatrix}$, $(1/10)\begin{pmatrix} -9 & 11 \\ -5 & 5 \end{pmatrix}$, $(1/5)\begin{pmatrix} -3 & 2 \\ 1 & 1 \end{pmatrix}$, $\mathbf{B}^{-1} = (1/2)\begin{pmatrix} 4 & 3 \\ 2 & 1 \end{pmatrix}$

4. $\begin{pmatrix} 16 & -4 & 6 \\ 3 & 3 & 4 \end{pmatrix}$

5. (a) $(1/26)\begin{pmatrix} 7 & -16 & 17 \\ 2 & 14 & -10 \\ -5 & 4 & -1 \end{pmatrix}$ (b) $x = 1, y = 1, z = -1$

6. (a) $\begin{pmatrix} -55 & -2 & -12 \\ 23 & 1 & 5 \\ -5 & 0 & -1 \end{pmatrix}$ (b) $x = -1, y = 2, z = 4$

7. $(1/16)\begin{pmatrix} -92 & 30 \\ -98 & -131 \\ -138 & -23 \end{pmatrix}$

9. For example, $\begin{pmatrix} 2/3 & -1/3 \\ 1/6 & 1/6 \\ 0 & 0 \end{pmatrix}$ with many other possibilities

10. (a) $(1/10)\begin{pmatrix} 16 & -20 & 22 & -8 \\ -21 & 25 & -27 & 13 \\ -25 & 35 & -35 & 15 \\ 24 & -30 & 28 & -12 \end{pmatrix}$

    (b) $(1/66)\begin{pmatrix} -47 & -12 & -14 & 43 \\ 51 & 6 & 18 & -27 \\ 14 & 12 & 14 & -10 \\ 25 & 12 & -8 & 1 \end{pmatrix}$

Section 5.9

1. $P = 6$
2. $P = 19/3$
3. $P = 18$
4. 100 instant, 400 regular
5. $P = 14$
6. $P = 700/3$
7. 25/7 jars of Econo-nut, 1250/7 jars of All-nutz, 425/14 jars of Lux-O-nuts, profit about $45.36
8. $6\frac{2}{3}$ portables, $28\frac{1}{3}$ Deluxe, no cheap models, price = $19,000

Section 5.10

2. minimum $P = 5$
3. minimum $P = 12$
4. minimum $P = 10/3$
5. minimum $P = -1/7$

Chapter 6

Section 6.1

1. Fall semesters: 1000, 1250, 1312.5; long-run, $1333\frac{1}{3}$. Spring semesters: 500, 625, 656.25; long-run, $666\frac{2}{3}$

2. $[(1/3)^k, 1 - (1/3)^k]$, $S = 1500$
3. Fall semesters: 1000, $1111\frac{1}{9}$, $1123\frac{37}{81}$; long-run, 1125. Spring semesters: $333\frac{1}{3}$, $370\frac{10}{27}$, $374\frac{118}{243}$; long-run, 375
4. ± 8
5. 4
6. $\pm 2/3$, 486, 97.2
7. $r = 1/7$; 6, 6/7, 6/49, 6/343
8. $a = 7 - 7\sqrt[5]{2/7}$, $r = \sqrt[5]{2/7}$
9. $x = \pm 12$, $y = 36$, $z = \pm 108$
10. 3, 7, 15, 31, 63, . . . , $2^{k-1} - 1$
11. 4, 6, 10, 18, 34, . . . , $2^k + 2$
12. 2, 1, -1, -2, -1, . . . , $u_6 = u_0$ and $u_7 = u_1$, and the cycle starts again.

Section 6.2

1. (a) 130/49 semesters; (b) 100/49; (c) 118/49; (d) 200/147
2. (a) (36/49, 13/49); (b) (39/49, 10/49); (c) (186/245, 59/245); (d) (127/147, 20/147)
3. From s_1, 495/56; from s_2, 215/28; from s_3, 55/7; $P(s_4 \mid s_1) = 99/112$, $P(s_5 \mid s_1) = 13/112$, $P(s_4 \mid s_2) = 43/56$, $P(s_5 \mid s_2) = 13/56$, $P(s_4 \mid s_3) = 11/14$, $P(s_5 \mid s_3) = 3/14$
4. Two-thirds of the time the process is absorbed in the absorbing state nearest the starting point. The expected time until absorption is two trials.
5. From the middle the probability of absorption is 1/2 in each state and the expected time until absorption is four trials. From position 2 or 4, the process is absorbed in the nearer absorbing state with probability 3/4, and the expected time until absorption is three trials.
6. Expected time until escape: from 1, 37/13 sec; from 2, 53/13 sec; from 3, 43/13 sec. Probability of exits: from 1, (5/13, 5/13, 3/13); from 2, (4/13, 4/13, 5/13); from 3, (3/13, 3/13, 7/13)
7. (a) Expected time until escape: from 1, 5 sec; from 2, 4 sec; from 3, 4 sec. Probability of exits: from 1, (1/2, 1/2); from 2, (5/8, 3/8); from 3; (3/8, 5/8)
   (b) Expected time until escape: from 1, 23/3 sec; from 2, 17/3 sec; from 3, 19/3 sec; from 4, 8 sec. Probability of exits: from 1, (10/21, 11/21); from 2, (13/21, 8/21); from 3, (8/21, 13/21); from 4, (3/7, 4/7)
8. (a) The process is absorbed in State 2 with probability 1. Starting from State 1, the expected time until absorption is 4/3 steps.
   (b) $\mathbf{N} = \begin{pmatrix} 2.4 & 0.8 \\ 0.8 & 1.6 \end{pmatrix}$, $\mathbf{B} = \begin{pmatrix} 1 \\ 1 \end{pmatrix}$
   (c) $\mathbf{N} = \begin{pmatrix} 1.6 & 0.8 \\ 0.8 & 2.4 \end{pmatrix}$, $\mathbf{B} = \begin{pmatrix} .4 & .6 \\ .2 & .8 \end{pmatrix}$

Section 6.3

1. 44, 3060
2. 1/2
3. $(a + b)/2$
4. $243.5 + (1/2)(3^{11})$
5. -10, 1005
6. 4, 3, 5550
7. 9990
8. $-26/33$
9. 0.18, 5/18
10. 615/154, 520/154, 425/154, 330/154, . . .

Section 6.4

3. (a) $2(n^2 + 1)$; (b) $9(2^{n-1}) - 4n - 2$
4. (a) $(2n^3 + 3n^2 - 5n - 30)/6$; (b) $[9n^2 + 12n - 7 + 61(-2)^n]/27$
5. (a) $(5^n + 3)/4$; (b) $(8^n - 1)/7(2^{n-1})$
6. (a) $[41(3^n) - 21(2^n) + 10^n]/7(2^n)$; (b) $3^n - 2^n + 2n + 4$; (c) $3^n + 2(4^n) + 2(-2)^n - 2$
7. $2000[(1.05)^n - 1]$, \$1257.79
8. $u_n = 1500[1 - (1/3)^n]$
9. $2000[5(3^n) + 12^n - 6(4^n)]/12^n$

Section 6.5

1. (a) $(3^n + 3)/2$; (b) $3(2^{n-1}) + (-2)^{n-1}$
2. (a) $[11(3^n) + 13 - 2n^2 - 4n]/8$; (b) $[25(2^n) + 11(-2)^n + 4(3^n)]/20$
3. (a) $[8 - 2(-5)^n]/3$; (b) $2(10^n) + [7(-5)^n - 13]/3$
4. (a) $u_n = 3(a^{n-1} - b^{n-1})/\sqrt{13}$, where $a = (1 + \sqrt{13})/2$ and $b = (1 - \sqrt{13})/2$
   (b) $u_n = 3(a^{n-1} - b^{n-1})/\sqrt{13} + 4^n$, where a and b are given above.
5. (a) $5(n + 1)(n + 2)/2$; (b) $3(8 - n)(1 + n)/2$
6. $u_n = (a^{n+1} - b^{n+1})/\sqrt{5}$, $a = (1 + \sqrt{5})/2$, $b = (1 - \sqrt{5})/2$
7. $u_n = (3 - 2n)3^{n-1}$

Chapter 7

Section 7.1

1. 132
2. 60
3. 160, 70
4. 1680, 2100, 1890
5. $14! \approx 8.7178 \times 10^{10}$
6. $6(13!) \approx 3.7362 \times 10^{10}$, $48(12!) \approx 2.992 \times 10^{10}$
7. $2 \cdot 6! \cdot 8! = 58{,}060{,}800$
8. $12! \approx 4.79 \times 10^8$; $11! = 39{,}916{,}800$; $5(11!) \approx 2 \times 10^8$; $20(10!) = 72{,}576{,}000$

9. 260,000
10. 1,300,000; 1,186,000
11. 1
12. 120
13. $2(10!)^2 \approx 2.6336 \times 10^{13}$

Section 7.2

1. $\binom{52}{7} \approx 1.3378 \times 10^8$
2. $4! = 24$
3. 200
4. $5! = 120$
5. 30
6. 210
7. (a) 5040; (b) 3780; (c) 50,400
8. 1764
9. 240
10. 525
11. 720
12. 51
13. 1260
14. (a) 90,720; (b) 1512; (c) 20,160

Section 7.3

1.

1
1 1
1 2 1
1 3 3 1
1 4 6 4 1
1 5 10 10 5 1
1 6 15 20 15 6 1
1 7 21 35 35 21 7 1
1 8 28 56 70 56 28 8 1
1 9 36 84 126 126 84 36 9 1
1 10 45 120 210 252 210 120 45 10 1
1 11 55 165 330 462 462 330 165 55 11 1
1 12 66 220 495 792 924 792 495 220 66 12 1

2. $x^{10} + 10x^9y + 45x^8y^2 + 120x^7y^3 + 210x^6y^4 + 252x^5y^5 + 210x^4y^6 + 120x^3y^7 + 45x^2y^8 + 10xy^9 + y^{10}$
3. $64x^6 - 96x^5y + 60x^4y^2 - 20x^3y^3 + (15/4)x^2y^4 - (3/8)xy^5 + (1/64)y^6$
4. $x^7 - 7x^5 + 21x^3 - 35x + 35x^{-1} - 21x^{-3} + 7x^{-5} - x^{-7}$
5. $7/9$
6. 20
7. -1792
8. $326{,}592x^5y^4$
9. $(15/4)x^2y^8$
10. 252
11. $x^8y^{-8} - 4x^5y^{-6} + 7x^2y^{-4} - 7x^{-1}y^{-2}$
12. $(35/8)y^8$
13. $70x^4$
14. $252a^{-5}x^{-5}$

Section 7.4

1. 1/6
2. 1/4165
3. $4\binom{13}{4}\binom{13}{3}^3 \Big/ \binom{52}{13} \approx 0.105$
4. (a) 10/33; (b) 15/28; (c) 28/55
5. (a) 579/2156; (b) 89/193
6. (a) 1/11; (b) 67/165; (c) 5/22
7. 2/5
8. 1/3, 8/15, 2/15, 4/5
9. 5/4
10. 94/143
11. 1/462
12. 6/5

Section 7.5

1. 0.2501, 0.4480, 0, 0.0003
2. (a) 0.1776; (b) 0.0003; (c) 0.9847; (d) 0.0835; (e) 0.0003
3. 2.8, 2.24
4. 4.8
5. 10.5, 0.2061, 0.2186, 11
6. 0.1792, 2.4, 1.44
7. 0.1296, 0.2074, 0.1659
8. 0.3125
9. 0.5520
10. 0.2637
11. 0.4480
12. First authority has a 0.0561 posterior probability of being right. Second authority has a 0.9439 posterior probability of being right.

Section 7.6

1. 0.5793, 0.9049, 0.0139, 0.0869
2. 0.5, 1.01, 1.52, −1.6, −1.43, −0.03
3. (a) 0.1525; (b) 0.8351; (c) 0.9770; (d) 0.9974
4. (a) 0.9032; (b) 0.1020; (c) 0.8686; (d) 0.0392
5. (a) 0.9733; (b) at least 0.75; (c) 0.9699
6. (a) 0.0796, 0.6318; (b) 0.0080, 1.0
7. 0.0796, 0.9534, 1.0
8. 0.9933, 0.8884
9. 0.0516
10. (a) 2.33; (b) 0.6554, 0.9251, 0.9934; 69.3, 66
11. 0.3557
12. The probability of 60 or fewer wins by Laver would be zero if the critic were correct. Therefore, he has poor judgment.

Chapter 8

Section 8.1

1. (a) $x = 3, y = 2$; (b) $x = 2, y = -4$
2. $x = \pm 2, y = \pm 1$
3. $(1/2, -11/2)$
4. $x = 0, -6$
5. $\mathbf{A}^{-1} = \begin{pmatrix} 2/11 & 3/11 \\ 1/11 & -4/11 \end{pmatrix}$; $\mathbf{B}^{-1} = \begin{pmatrix} -1/3 & -2/3 \\ 4/3 & 5/3 \end{pmatrix}$; $\mathbf{C}^{-1} = \begin{pmatrix} 1/6 & 2/3 \\ 0 & 1 \end{pmatrix}$
6. (a) $x = 5/18, y = -5/2$; (b) No solution

Section 8.2

1. -18
2. 0
3. 48
4. 0
5. -11
6. 23
7. 82
8. 6
9. 7
10. 0
11. ± 3
12. $-1, -3$
13. $-7.5, -3$

Section 8.3

1. $\mathbf{A}^{-1}$ exists, $\mathbf{B}^{-1}$ does not exist, $\mathbf{C}^{-1}$ exists, $\mathbf{D}^{-1}$ exists if $b \neq a$.
2. $\det(\ \ \mathbf{C}) = -294 = 7(-42)$
3. $x = 2/9, y = 41/27, z = 10/27$
4. $x = 2, y = -3, z = 1$
5. $x = -1, y = -2, z = 2$
6. The corresponding operations are performed on the columns of **A**.
7. $\mathbf{E}_2^{-1} = \begin{pmatrix} 1 & 0 & 0 \\ 0 & 1 & 0 \\ 0 & 0 & -1/2 \end{pmatrix}$, $\mathbf{E}_3^{-1} = \begin{pmatrix} 1 & 0 & 0 \\ 0 & 1 & 0 \\ 4 & 0 & 1 \end{pmatrix}$

10. $\det(\mathbf{A} + \mathbf{C}) = 78$
11. There are many possibilities; one of the easiest is

$$\mathbf{A} = \begin{pmatrix} 1 & 0 & 0 \\ 0 & 1 & 0 \\ 0 & 0 & 0 \end{pmatrix}, \mathbf{B} = \begin{pmatrix} 0 & 0 & 0 \\ 0 & 0 & 0 \\ 0 & 0 & 1 \end{pmatrix}, \mathbf{A} + \mathbf{B} = \mathbf{I}$$

12. $-1, 3/5$
17. For example, let $\mathbf{A} = \mathbf{E}$.
18. $(1 + b)(1 - b), (1 + 2b)(1 - b)^2, \ldots, [1 + (n - 1)b](1 - b)^{n-1}$, for an $n \times n$ matrix.

Index

INDEX